NONSKID BASICS: CONTRACTOR'S GUIDE IN MILITARY APPLICATION

By
Johnny M. Sanchez

AMPP™

AMPP
Association for Materials Protection and Performance

PRINT ISBN: 978-1-57590-493-1
E-book ISBN: 978-1-57590-494-8

AMPP
Association for Materials Protection and Performance
15835 Park Ten Place
Houston, TX 77084
ampp.org

Contents

Acknowledgments

I would like to thank my beautiful wife, Karen M. Sanchez, for supporting me in the writing of this book; her patience was key for me writing and finishing this third book of mine.

I would not be who I am in the industry if it wasn't for the guidance and support of numerous people:

> I thank my mentor and nonskid "pops"—Bernard Meartz—for always being there for me. Cad Parco is my "brother in nonskid"—we've worked together for 30 years on some of the toughest nonskid jobs for the Navy. Gordon Kulgien has been my AMPP co-instructor who many, many years ago supported me becoming an SSPC/AMPP Lead Instructor. Ramon Jauregui, from SWRMC Engineering, has always supported me in the industry. Dave Schoales is a dear friend and nonskid colleague. We bounce ideas off each other and together, we find ways toward better nonskid application. I would also like to thank Andre Lomeli for helping fix some of my image art for this book.

Lastly, I am extremely proud of two of my protegés—Felix Perez and Tomas Isydro—for their careers and their input in the nonskid industry...keep up the outstanding work!

Dedication

"Nonskid Basics" is dedicated to two colleagues in nonskid who are no longer with us. They both honorably served our country in the U.S. Navy for more than 20 years before retiring. They then went on to start the Commander, Naval Air Forces Pacific Fleet (COMNAVAIRPAC) Nonskid Onsite Representative (OSR) program, as the first nonskid onsite representatives on CVN[1] aircraft carrier flight decks. I had the privilege of working with these men who became dear friends:

Bud Cothern and Michael Edward Stacks

1. Carrier, Volpane, Nuclear

About the Author

Johnny M. Sanchez started his path in nonskid coatings at the Long Beach Naval Shipyard in 1993, working for a small nonskid coatings company called Bay Decking. After President Clinton closed down multiple military bases in 1994—including the Long Beach Naval base—Johnny moved to San Diego, where he flourished in the coatings industry working for Surface Technologies. Starting as a nonskid laborer applicator, he progressed to painter, then blaster, then foreman, to finally becoming a coatings inspector. The Association for Materials Protection and Performance (AMPP, formerly NACE International, and the Society for Protective Coatings [SSPC]) has played a major role in Johnny's growth as a coatings inspector. He is an AMPP lead instructor for teaching courses from the Naval Sea Systems Command (NAVSEA) Basic Paint Inspector (NBPI) Certification Program including as an instructor for the Coatings Application Specialist (CAS) and the Blaster Certification

(C-7) programs. He is now the owner and operator of Prime Time Coatings, Inc., a nonskid coating company.

Johnny is the recipient of numerous recognitions and awards including one from the USS *Nimitz* CVN-68[2] in 2008, the USS *Carl Vinson* CVN-70 in 2016, and the winner of *Coatings Pro* Inaugural Most Valuable Player (MVP) in 2017. Johnny is most known for his role as a nonskid coatings subject matter expert.

2. Carrier, Volpane, Nuclear

Editor's Notes

Legacy figures and photographs are included in this book, which have quality issues because of the technology available at the time they were made or taken. Where possible, efforts have been made to enhance the quality of these images using modern technology. Photographs are specifically credited when obtained from sources other than the author.

In this book, the nonstandard use of certain abbreviations—such as those for square feet, cubic feet per minute, square feet per gallon, etc., are styled as "sqft", "CFM", and "sqft per gallon", respectively, instead of the standards of ft^2, $ft^3\ min^{-1}$ or ft^3/min, and $ft^2\ gal^{-1}$ or ft^2/gal. The author specifically used these to reflect what is used in the nonskid industry with U.S. Navy applications.

Certain U.S. Government (and other) standards are directly quoted with their original capitalization, emphasis, and actual ways of stating things. When the reader is directed to specific sections or paragraphs, they are being directed to those items in the standards themselves, and not places in this book, unless otherwise directed by a phrase within square brackets and a change in the font. Paragraphs and such that are direct quotes from these standards are in a different font from the rest of the text.

Because this book only references standards from various organizations and the U.S. Government, the usual-and-customary way AMPP cites and formats reference information (i.e., chronologically as cited in a given chapter or the book as a whole) has been suspended. Instead, a bespoke system that is better suited for exclusively listing standards is used. All standards are listed in the back of the book first alphabetically by organization and then sequentially. For example, all ASTM standards are listed sequentially before MIL-PRF standards, NACE standards appear before SSPC standards, etc. Note that because NACE International and SSPC merged in 2021—forming AMPP—in the References, all NACE, SSPC, and joint NACE/SSPC and SSPC/NACE, standards are listed under "AMPP". Likewise, joint ANSI/NACE standards are also listed under AMPP.

Nonslip deck coatings, rapid-cure, and slip-resistant coating systems are all interchangeable terms, however, the nomenclature used by the U.S. Navy for these is "nonskid".

Some concepts and phrases are repeated throughout the book, where applicable, for clarity and completeness.

Foreword

The manufacturers' requirements for this aggressive high-tech safety feature must meet the conditions of numerous standards ***and be approved by*** the Navy.[3] These Navy-approved standards give the required ranges by way of profile depth, wet film and dry film thickness minimums and maximums, allowable application ambient conditions, and also for recoating windows for a particular coating to ensure maximum coating-life longevity.

Because these types of products are used on military vessels, contractors are under contractual obligations and requirements to meet the Naval Sea Systems Command (NAVSEA) Standard Item 009-32,[4] which covers all coatings maintenance onboard military vessels. *Note that NAVSEA Standard Item 009-32 is considered the governing document, which supersedes all manufacturer requirements.* If NAVSEA Standard Item 009-32 has a more stringent requirement, it is to be followed in lieu of the ASTM International ASTM F718-22.[5] All manufacturers' ASTM F718-22 data sheets must be reviewed by NAVSEA and have the watermark "NAVSEA REVIEWED" printed on them. This NAVSEA-reviewed manufacturers' ASTM F718-22 data sheet can be found on the Naval Surface Treatment Center (NST) website.

Nonskid coatings are a complex coating system containing aggregates, which help provide a slip-resistant deck on walk areas or vehicle traffic areas, and most importantly—on aviation decks. This nonskid coating is a major safety feature onboard U.S. Navy vessels.

The most common coating removal process 25–30 years ago was through centrifugal blasting (walk-behind blasting) meeting the requirement of SSPC-SP-10[6,7], which created debris and dust contamination. Throughout the years there have been many technical advances in the removal process, which have helped both with environmental concerns and most importantly by reducing the amount of exposure to the personnel. This process is accomplished through the use of ultra-high

3. This process is through the Naval Sea Systems Command (NAVSEA) Qualified Product Data List (QPD) identified under Military Performance (MIL-PRF)-24667) and the manufacturers' NAVSEA-reviewed ASTM-718.
4. NAVSEA Standard Item 009-32, "Cleaning and Painting Requirements; accomplish"
5. ASTM F718-2, "Standard Specification for Shipbuilders and Marine Paints and Coatings Product/Procedure Data Sheet"
6. SSPC, formerly the Society for Protective Coatings
7. SSPC-SP 10/NACE No. 2-2006, "Near-White Metal Blast Cleaning"

pressure water jetting with pressures ranging from 35,000–50,000 psi, meeting the standard of SSPC-WJ-2L[8] (previously SSPC-SP-12).

Most rapid-cure slip-resistant coatings are abrasive, containing

- 40–50% (on average) aluminum oxide
- 20–35% barium sulfate
- 10–20% epoxy resin

Nonskid coatings are resistant to most acids, alkalis, solvents, grease, detergents, gasoline, jet fuels, and other hydraulic fluids. There are also nonabrasive slip resistant coatings where the aggregate is glass, cork, and/or aluminum. Aluminum aggregates are most commonly used in naval nonskid coatings for specific areas where a softer aggregate is needed.

8. SSPC-SP WJ-2/NACE WJ-2 "Waterjet Cleaning of Metals–Very Thorough Cleaning" Joint Surface Preparation Standard

1 Types of Surface Preparation

1.1 Surface Cleanliness Prior to Nonskid Removal

Surface cleanliness is essential in any preservation process. Therefore, ensuring the surface cleanliness is the critical first step prior to startup with any equipment that uses tools or abrasives affecting the surface at high velocities to remove existing coating or nonskid.

Cleanliness is vital because any tools or abrasives that hit the surface will embed whatever is on the existing nonskid surface into the substrate, as shown in Figure 1.1. Nonskid systems are usually used in high traffic areas, and these areas are contaminated with multiple contaminants ranging from oil, grease, aqueous film forming foam, to petroleum. If the surface preparation were to take place without ensuring these contaminants are removed, then the contamination will result in premature nonskid failure.

The standard set forth in U.S. Naval Sea Systems Command (NAVSEA) Standard Item 009-32 is to accomplish cleanliness in accordance with SSPC-SP 1, "Solvent Cleaning."

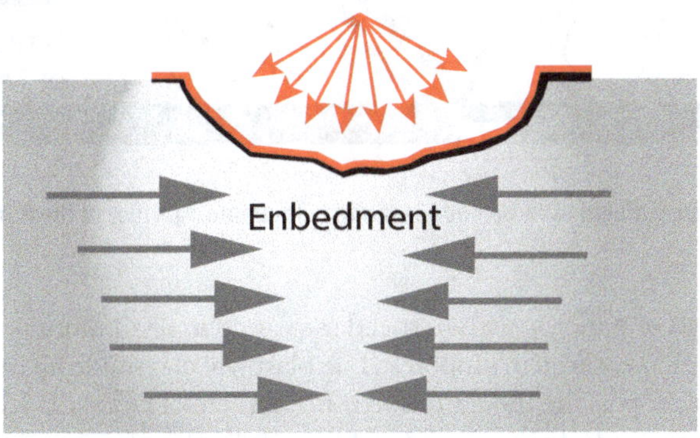

FIGURE 1.1 Abrasive blast contaminants embedment.
Original image courtesy of Blast Track Inc., with embedment line made by the author.

1.1.1 SSPC-SP 1 Definition

According to the standard in SSPC-SP 1, "When viewed without magnification, a solvent-cleaned surface shall be free of visible oil, grease, dust, dirt, drawing and cutting compounds, and other visible soluble contaminants. "Visible" means detectable with normal or corrected normal vision without the use of additional test equipment."

1.2 Centrifugal Blasting

Nonskid coatings are normally applied on decks. The equipment used are walk-behind portable centrifugal blasters. Abrasive blasting is the operation of forcibly propelling a stream of abrasive material against a surface. The objective is to make a roughened surface for the anticorrosive barrier coat (primer) to adhere to the surface being coated; a process known as "mechanical adhesion". This roughened surface can be identified as an "anchor tooth" or "profile" rendering a peak-to-valley surface. *For the purposes of this guide, hereafter, the roughened surface is identified as the "profile".* A centrifugal wheel (Figure 1.2) is used to propel the blasting material (often called the "media").[1]

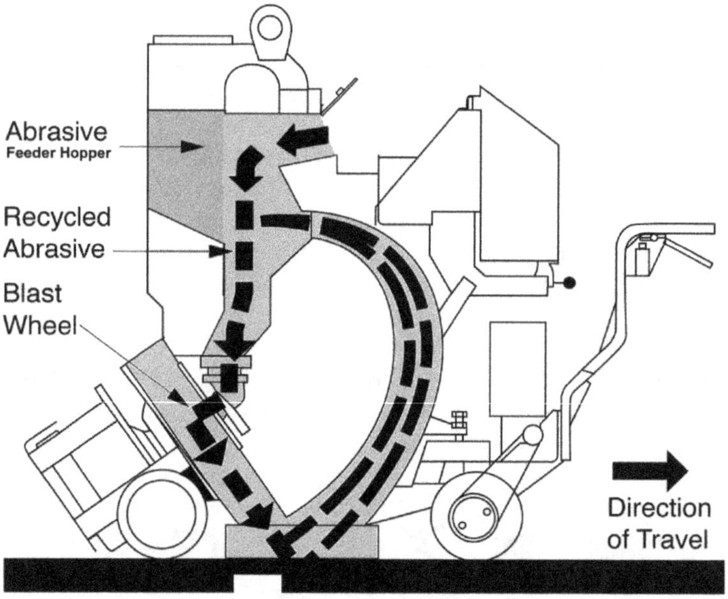

FIGURE 1.2 Centrifugal walk-behind blast machine. Photo courtesy of Blast Track Inc.

Centrifugal blasting consists of a motor, with its speed measured in revolutions per minute (rpm), which in turn spins the blast wheel, measured in amperes. The blast wheel is made up of interchangeable steel blades (Figure 1.3), which propels the media towards the surface. The media is fed through a hopper from which the operator can control the amount of blast media that drops to the blast wheel. This is a closed loop recycling system that uses spherical blast media referred to as "shot".

1. The first abrasive blasting process was patented by Benjamin Chew Tilghman on October 18, 1870. https://coatingsystems.com/how-does-sandblasting-work/

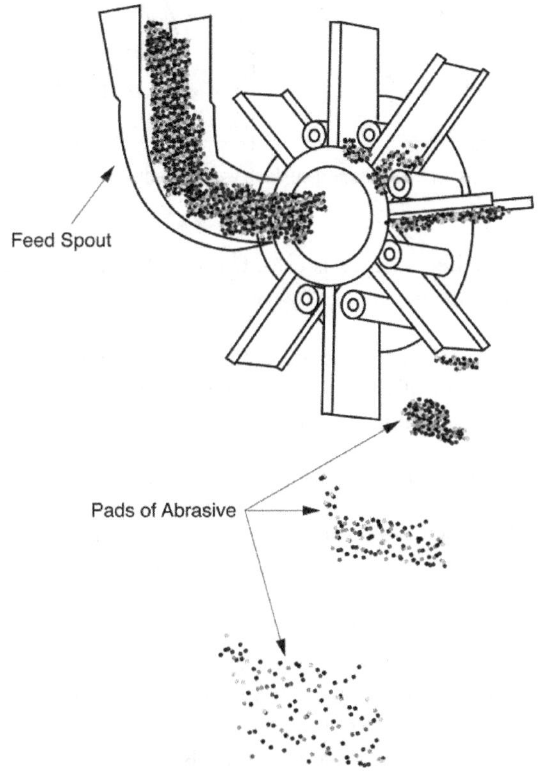

FIGURE 1.3 Interchangeable blades that propel the abrasive media towards the deck. Photo courtesy of Blast Track Inc.

1.2.1 Centrifugal Abrasive Blasting vs. Nozzle Abrasive Blasting

Centrifugal abrasive blasting is unlike nozzle abrasive blasting. The nozzle abrasive blasting process uses high-pressured air to send a stream of blast media to the surface. The high-pressured air used is read in pounds per square inch (PSI). This type of blasting uses an angular type of blast media, which does not rebound like the rounded shot blast media. The angular, jagged shape of the blast media does not make it centrifugal blast friendly or economical on large, horizontal flat surfaces, like flight decks. Because it breaks down much faster, it loses its recyclability and does not have the rebound factor, which is needed in portable centrifugal blasting systems.

NOTE: There are differences in the profile when using the round spherical blast media vs. the angular jagged grit blast media (Figure 1.4).

- Shot media results in a peened surface.
- Grit media results in an etched surface.

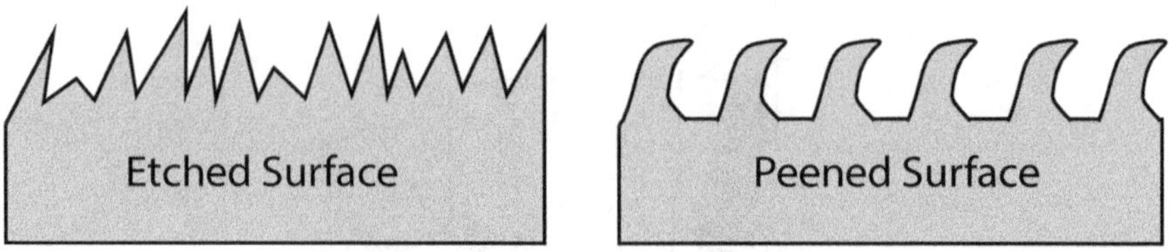

FIGURE 1.4 Etched surface vs. peened surface.

In the coatings discipline, an etched surface is the ideal profile for coatings adhesion. Open-nozzle blasting using angular grit media is not normally used in nonskid operations.

Most nonskid coatings are used on large flat areas. This, combined with the thickness of the nonskid coating, makes it difficult to accomplish nonskid removal for surface preparation by open-nozzle blasting using air pressure, even at maximum PSI.

Open-nozzle blasting is not used on most nonskid projects because of the following considerations:

- Closed loop centrifugal blasting does not require 100% containment, unlike open-nozzle blasting.
- Open-nozzle blasting uses angular grit media, which breaks down faster than spherical shot media used in centrifugal blasting operations.
  - Open-nozzle blasting using grit blast media is not economical on large open areas because of constantly having to fill the pressurized pot[2] with new media, which ends up as waste once used. This is costly not only by incurring cost for procuring large amounts of grit blast media, but also for disposing used blast media, which normally never gets recycled.
- Spherical shot media is more recyclable and rebounds back through the closed loop centrifugal blast system for minimal blast media loss.
- Nonskid, on average, is 3/8" thick.
- Using nozzle blasting at 120 PSI of air pressure would take 20 times longer to remove existing nonskid at its average thickness. Nozzle blasting is operated by personnel so the production rate would drop because of operator fatigue.
- Centrifugal blasting is machine operated; it uses a motor read in revolutions per minute, which spins the blast wheel steel blades. This in turn propels the shot blast media towards the surface, providing a quicker and more efficient way of removing the thick, aggregated nonskid coating.
- For nonskid coatings, there are two locations that get open-nozzle blasting. These locations will be covered more in depth in Chapters 2 and 5.
  - Vertical Launch System (VLS)—because of its design, it cannot be centrifugal blasted.
  - Areas that receive Thermal Sprayed Nonskid (TSN)—the U.S. Naval Research Laboratory (NRL) has done tedious testing on the best mechanical adhesion for TSN. Because nozzle blasting creates an etched surface, NRL has concluded this type of profile proved to have the best adhesion and will give the thermal spray nonskid the longevity needed for its intended life service.

Per NAVSEA Standard Item 009-32, the required standard to achieve when centrifugal blasting on military naval vessels is to meet the Society for Protective Coatings (formerly the Steel Structures Painting Council) SSPC-SP 10.

2. The use of pressurized pots is a requirement of the American Society of Mechanical Engineers (ASME). The pot must be able to meet a certain degree of pressure to ensure the pot can operate with grit in it at air pressure read in pounds per square inch (PSI).

1.2.2 SSPC-SP 10 Definition

This standard, "Near-White Metal Blast Cleaning," relates to accomplishing centrifugal blasting (Figure 1.5) and gives the following requirements (quoted from the standard):

- A near-white metal blast cleaned surface, when viewed without magnification, shall[(3)] be free of all visible oil, greases, dust, dirt, mill scale, rust, coating, oxides, corrosion products, and other foreign matter, except for staining.
- Random staining may consist of light shadows, slight streaks, or minor discolorations caused by stains of rust, stains of mill scale, or stains of previously applied coating.
- Random staining shall be limited to no more than 5% of each surface unit area. Area of determination shall be 3 in × 3 in (9 in^2) or 76 mm × 76 mm (5776 mm^2).

FIGURE 1.5 The USS *Green Bay* LPD-20 flight deck surface preparation to an SSPC-SP 10. Accomplished in April 2012.

SSPC-VIS 1 is a visual guide (comparater) for abrasive blasted surfaces (Figure 1.6) made up of photos for each standard. These photos have been taken of abrasive blasted surfaces depicting four rust grades of carbon steel. The purpose of SSPC-VIS guides is to establish a visual criteria to meet in case of a difference of opinion "visually" on what the standard should look like in a real-life scenario.

NOTE: The VIS 1 guideline photos were taken in a lab setting under optimum conditions.

3. The use of "shall" in standards means that the action is mandatory, not optional.

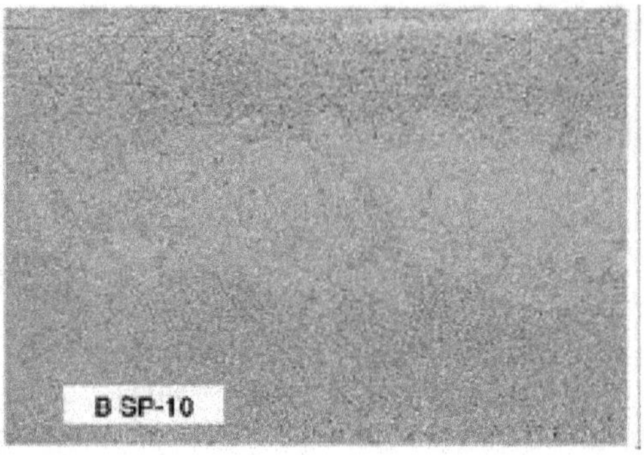

FIGURE 1.6 Photo of abrasive blasted suface to SSPC-SP 10 standard.

1.2.3 SSPC-VIS 1

This comprehensive reference, "Guide and Reference Photographs for Steel Surfaces Prepared by Dry Abrasive Blast Cleaning," visually represents the following standards:

- SSPC-SP 7, "Brush Off Blast Cleaning," tightly adhered scale, rust, and coatings allowed
- SSPC-SP 6, "Commercial Blast Cleaning," 33% staining allowed
- SSPC-SP 10, "Near-White Metal Blast Cleaning," 5% staining allowed
- SSPC-SP 5, "White Metal Blast Cleaning," no staining allowed
- SSPC-SP 14, "Industrial Blast Cleaning," 10% staining allowed

Rust grades identified in SSPC-VIS 1:

- A—Steel surface completely covered with adherent mill scale; little or no rust visible
- B—Steel covered with both mill scale and rust
- C—Steel surface completely covered with rust; little or no pitting visible
- D—Steel surface completely covered with rust; pitting visible

1.2.4 SSPC-SP 17

Another standard involving centrifugal abrasive blasting is SSPC-SP 17, "Thorough Abrasive Blast Cleaning of Non-Ferrous Metals." It is used for nonskid systems on Navy ships, usually with the aluminum alloy (Al) H-116 of the 5000 series, as seen in Figures 1.7 and 1.8.

FIGURE 1.7 Photo of CVN-70 Elevator #1 (ACE-1). Because the elevator is aluminum, the required abrasive standard is SSPC-SP 17 for nonferrous metals.

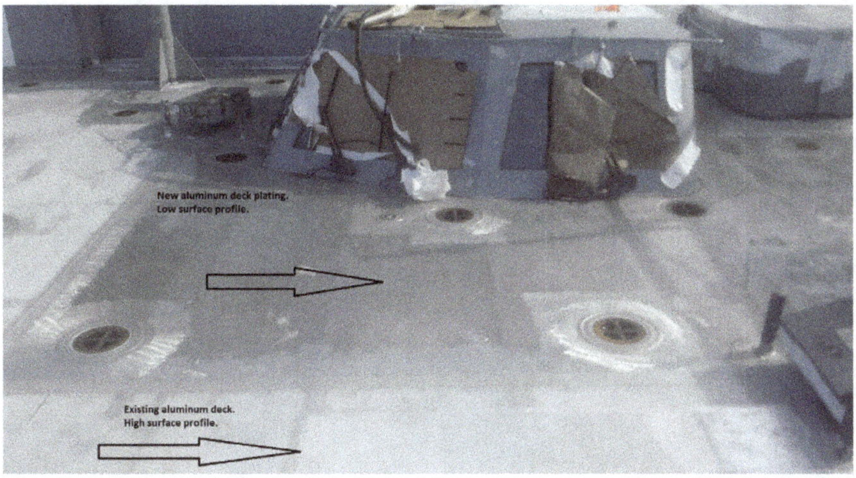

FIGURE 1.8 Ultra high-pressure water jetting (UHPWJ) is usually the surface preparation done for aluminum surfaces on naval vessels. When the profile is nonexistent because of repairs or it is too high, the repair activity shall accomplish centrifugal abrasive blasting to implement the required profile in accordance with NAVSEA Standard Item 009-32.

The following class ships have aluminum decks:

- Guided Missile Cruisers (CG Class) flight decks
- Aircraft Carrier (CVN[4]) Elevators (ACE)-1 through ACE-4
- Ships Assault Craft Unit (ACU)-5 Landing Craft Air Cushion (LCAC)—the entire craft is Al (aluminum)
- Littoral Combat Ship (LCS) of the Independence and Freedom classes—the entire craft is Al (aluminum)

1.2.4.1 SSPC-SP 17 Definition[5]

- Thorough abrasive blast cleaned surface: A thorough abrasive blast cleaned surface, when viewed without magnification, shall be free of all visible oil, grease, dust, dirt, oxides (corrosion products), coatings, and other foreign matter. Random color variations shall be limited to no more than 5% of each unit area of surface (approximately [~] 5,800 mm^2 [9.0 in^2] (i.e., a square 76 mm × 76 mm [3.0 in × 3.0 in], and may consist of light shadows, slight streaks, or minor discolorations. In SSPC-SP 17, Section A6 of Appendix A provides additional information about color variation. Section A7 of Appendix A provides an example of a specification statement.
- The entire metal surface shall be subjected to the abrasive blast to achieve the specified degree of cleaning and to produce a dense and uniform surface profile (e.g., depth and texture) on the bare metal substrate. The peaks and valleys on the surface shall form a continuous pattern, leaving no smooth, unprofiled areas.
- Immediately prior to coating application, the entire surface shall comply with the degree of cleaning as specified herein. Section A8 of Appendix A provides additional information regarding time from blast until coating application.

1.2.4.2 Steel Shot and Other Abrasive Blasting

In February 2013, a NAVSEA message identified as "R-111724Z" went out fleet wide, stating the use of steel shot or any abrasive blasting was prohibited on the LHA, LHD, LPD, and LSD classes of ships. Because of foreign object damage (FOD) from debris, it was found that steel shot was the cause of "fodding" an aircraft engine (Figure 1.9). The Navy took immediate action to protect military aircraft that range in cost between $35,000,000 and $45,000,000.

4. CVN= "Carrier, Volplane, Nuclear"—a U.S. Navy hull classification symbol for nuclear aircraft carriers, Wikipedia source.
5. As with other standards' definitions quoted, these bullets are direct quotes from SSPC-SP 17. Note that the sections and appendices given reference parts of the actual standard and are not found in this book.

FIGURE 1.9 Photo of an F35B aircraft on an LHD flightdeck. It shows why steel shot cannot be used on this class of ship because of foreign object debris (FOD) damage: FOD can get sucked into a jet engine and may cause it to crash. Photo courtesy of Lt. Commander David Martinez.

The U.S. Navy has put in place contingencies for using steel shot on these classes of ship, if absolutely necessary. Per the NAVSEA Standard Item 009-32 message, "any blast media containing steel abrasives shall not be used without *approval* from the applicable *type commander* via a major departure from specification (DFS). This DFS shall be submitted to NAVSEA prior to commencement of work. All work shall stop until major DFS is approved."

Government and contractor personnel have identified reasons for the use of abrasive blasting. Because ultra high-pressure water jetting (UHPWJ) is deemed the preferred method of nonskid removal meeting SSPC-WJ-2L (where the "L" indicates "light cleanliness"), per the NAVSEA message. SSPC-WJ-2L will be covered in more detail later in this chapter. UHPWJ does not create a profile. Gauge readings are taken to ensure the required profile depth is met. If the profile is out of range or nonexistent, then a DFS will be submitted to gain approval for the use of abrasives to implement the required profile set forth in NAVSEA Standard Item 009-32.

There are extreme situations in which the Navy will allow deviations from the specification, such as to support escalating issues in other regions of the world. The two photos below were taken by the author when he was the nonskid inspector on the USS *Nimitz* CVN-68 (Figure 1.10), in 2003. Surface Technologies was tasked to remove and replace the nonskid in the arresting gear area. Although one would usually never see working military aircraft near abrasive blasting operations, with a short deadline of eight days to meet the scheduled departure from San Diego, the USS *Nimitz* was loaded with aircraft preparing for its mission in Iraq. COMNAVAIRPAC[6] saw the need to redo the nonskid in the wire area for the safety of the ship's personnel during flight operations because the landing area on an aircraft carrier area is the most critical part of the flight deck.

The Limited Nonabrasive Nonskid is expected to meet 10,000–15,000 traps (arrests) depending on the type of nonskid used. Type-1 Comp-L[7] nonskid is identified to meet 10,000 traps. Type-5 Comp-L nonskid is identified to meet 15,000 traps.

6. COMNAVAIRPAC: Commander, Naval Air Force, U.S. Pacific Fleet
7. More detail about the various types of nonskid is given later in the book.

FIGURE 1.10 (top) USS *Nimitz* CVN-68 with the loaded aircraft surrounding abrasive blasting operations; (bottom) wire box landing area after nonskid operations were completed (2003).

1.3 Ultra High-Pressure Water Jetting (UHPWJ)

High-pressure water jetting is not only economical, but with so many environmental advantages, it has become the preferred process for nonskid removal by the Navy.

UHPWJ (Figure 1.11) is defined by AMPP (formerly SSPC and NACE) as cleaning with water pressures above (25,000 PSI). The basic UHPWJ system consists of a high-pressure pump, hoses, and large, high, cubic feet per minute (CFM) vacuums. The high-pressure hoses used must have a bursting strength 2.5 times the maximum-rated operating capacity. Therefore, a 1,700 bar (25,000 PSI) unit requires high-pressure hoses with a minimum bursting strength of 4,250 bar (62,500 PSI). The operating systems used in field surface preparation consist of a control valve assembly spray bar, which provides the ultra high-pressure water jets through which are diamond cut orifices or tips that have very small openings. These systems can have a single jet, a fan-shaped jet, or multiple rotating jets. The most common arrangement on UHPWJ equipment for removing nonskid from steel is a rotating 16-inch bar with multiple tips. A single jet has a very small blast pattern. Multiple jets on a rotating 16-inch bar increase the size of the blast pattern to increase productivity in nonskid removal. Rotation also increases the cutting action. With UHPWJ, the pressure, i.e., velocity, of the water is the main energy characteristic that performs the nonskid removal.

Flow rate plays a role equal to water velocity in cleaning. Pressure falls off quickly with distance. Productive nonskid removal with UHPWJ requires keeping the blast bar with tips 0.25–0.5 inches from the surface.

FIGURE 1.11 Photo of UHPWJ nonskid removal.

1.3.1 Advantages of UHPWJ vs. Centrifugal Abrasive Blasting

The advantages of UHPWJ, when compared to centrifugal abrasive blasting, include the following:

- UHPWJ does not create dust emissions.
- All waste generated in UHPWJ operations is contained in large vacuum boxes, which is then transferred to a Baker tank or Riser after testing the wastewater.
- No abrasives are used, eliminating any FOD issues from nonskid operations.
- Pump pressures can be lowered to accomplish cleaning of nonskid.
- For UHPWJ, the SSPC-SP 1 requirement is waived. The contractor has six hours to cover bare steel with primer, or the SSPC-SP 1 shall be reverified by accomplishing the dust tape test found in ISO-8502-3.
- One of its key advantages is the removal of chlorides from the substrate, lowering the conductivity across the metallic pathway (steel deck). Lowering the chlorides helps mitigate the corrosion process under the nonskid system, preventing premature coatings failure in the form of undercutting rust, thereby extending the life of the nonskid coating's system (Figure 1.12).

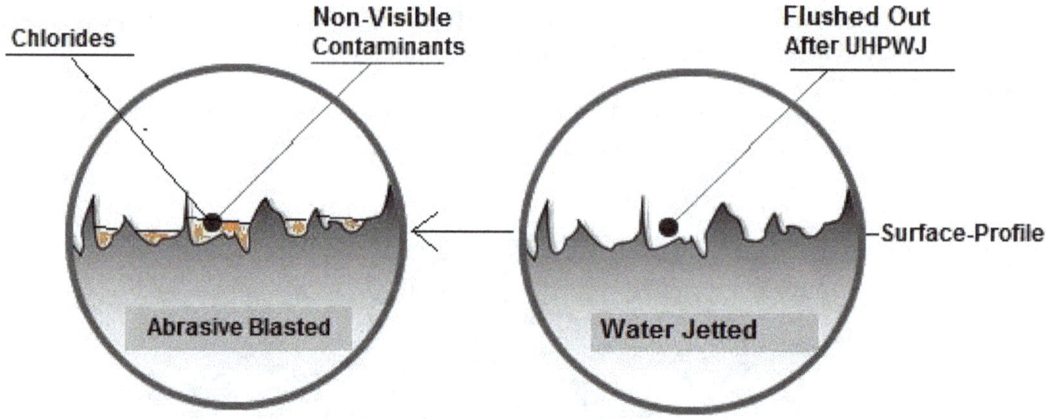

FIGURE 1.12 Example of nonvisual contaminants before and after UHPWJ.

1.3.2 Disadvantages of UHPWJ vs. Centrifugal Abrasive Blasting

The disadvantages of UHPWJ, when compared to centrifugal abrasive blasting, include the following:

- UHPWJ is not normally used in new construction because it is not an effective way to remove mill scale.
- UHPWJ is not used on the mine counter measure (MCM) class ship (a wooden ship with glass reinforced plastic [GRP]) because of the high potential for damage.
- UHPWJ does not create a surface profile, it can only expose existing profile.

As can be seen in the list above, the benefits of UHPWJ outweigh the disadvantages.

Per NAVSEA Standard Item 009-32, the required standard to achieve when using UHPWJ on military naval vessels is to meet that of SSPC-WJ-2L as seen in Figure 1.13.

FIGURE 1.13 The blasted surface meeting SSPC-WJ-2L.

1.3.3 SSPC-WJ-2L Definition

Below are the applicable paragraph numbers from this standard (quoted from the standard).

- 2.1—Very Thorough Cleaning (WJ-2): A metal surface after Very Thorough Cleaning, when viewed without magnification, shall have a matte (dull, mottled) finish and shall be free of all visible oil, grease, dirt, rust, and other corrosion products except for randomly dispersed stains of rust and other corrosion products, tightly adherent thin coatings, and other tightly adherent foreign matter. The staining or tightly adherent matter shall be limited to no more than 5 percent of each unit area of surface and may consist of randomly dispersed stains of rust and other corrosion products or previously applied coating, tightly adherent thin coatings, and other tightly adherent foreign matter.
- 2.1.1—A unit area of surface is an area approximately 5,800 mm^2 [9.0 in^2] (i.e., a square 76 mm × 76 mm [3.0 in × 3.0 in].
- 2.1.2—Coatings, mill scale, and foreign matter are considered tightly adherent if they cannot be removed by lifting with a dull putty knife. (Paragraphs A4 and A5 of Appendix A provide additional information.)
- 2.1.3—The gray to brown-black discoloration remaining on corroded and pitted carbon steel that cannot be removed by further waterjet cleaning is not considered part of the percentage staining.
- Light (L) flash rusted surface: A carbon steel surface that, when viewed without magnification, exhibits small quantities of a rust layer through which the carbon steel substrate may be observed. The rust or discoloration may be evenly distributed or present in patches, but it is tightly adherent and not easily removed by lightly wiping with a cloth.

1.3.4 SSPC-VIS 4

This is a visual comparater for abrasive blasted surfaces, made up of photos for each standard depicting four rust grades of carbon steel.

NOTE: The VIS 4 guideline photos were taken in a lab setting under optimum conditions.

The standards visually represented in SSPC-VIS 4 are listed below with their respective levels of cleaning defined:

- SSPC-WJ-1, "Waterjet Cleaning of Metals—Clean to Bare Substrate"—No staining allowed
- SSPC-WJ-2, "Waterjet Cleaning of Metals—Very Thorough Cleaning"—5% staining allowed
- SSPC-WJ-3 "Waterjet Cleaning of Metals—Thorough Cleaning"—33% staining allowed
- SSPC-WJ-4 "Waterjet Cleaning of Metals—Light Cleaning"—Tightly adhered scale, rust and coatings allowed

Rust grades identified in SSPC-VIS 4:

- A—Steel surface completly covered with adherent mill scale; little or no rust visible. No photo
- B—Steel covered with both mill scale and rust
- C—Steel surface completely covered with rust; little or no pitting visible
- D—Steel surface completely covered with rust; pitting

1.3.4.1 Open-Wand UHPWJ vs. Closed-Loop Vacuum UHPWJ

Open-wand UHPWJ (Figure 1.14) is not contained or connected to a vacuum where all the water used is collected and the steel being blasted is automatically exposed to oxygen. Because of the large volume of water and oxygen, the steel surface automatically rusts ("flashes") in the form of red rust. It is more difficult to meet the SSPC-WJ-2L standard using open wand UHPWJ.

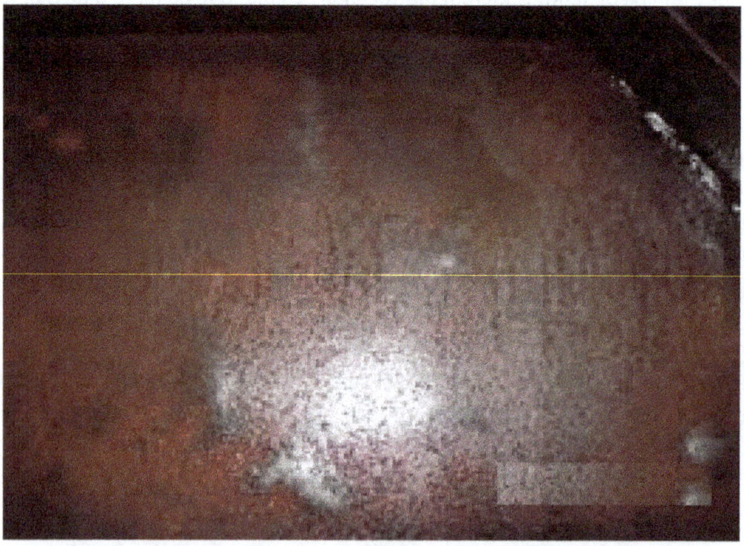

FIGURE 1.14 Open-wand UHPWJ showing red rust.

Red Rust: Rust from Iron (III) oxides form due to high oxygen and water exposure resulting in red rust, known as hydrated oxide $Fe_2O_3 \cdot H_2O$. Red rust is the result of heavy exposure to air and moisture; it is usually out in atmospheric conditions. One feature of red rust is that there is uniform corrosion.

A closed-loop vacuum UHPWJ is connected to a large vacuum with an extremely strong cubic feet per minute (CFM). The hydro deck mower (the machine that does the ultra high-pressure water jetting) has a rubber seal around its base, which when combined with the vacuum's CFM, creates an oxygen deficient atmosphere as the UHPWJ passes over the deck. This results in "black rust" in areas that were previously heavily corroded (Figure 1.15).

FIGURE 1.15 Closed-loop vacuum surface preparation according to SSPC-WJ-2L, where black rust is shown.

Black rust is rust from iron (III) oxide, i.e., Fe_3O_4. It forms in limited oxygen and low moisture conditions when combined with iron (II) oxide.

Black rust ("magnetite") can be visually identified as a thin, black film, which is the result of oxidation in a low oxygen environment. It has the appearance of having almost a black stain. The areas exhibiting the black rust most likely had something covering them, which prevented oxygen from reaching the surface. This type of rust is a more stable rust layer that does not propagate as rapidly as other rust forms. These areas of black rust are porous in nature and need additional paint to meet the required dry film coverage.

When accomplishing UHPWJ, the contractor must have or acquire a large amount of equipment and be able to make logistically smart movements on large projects. Knowing how to mobilize a flight deck is crucial when working with the U.S. Military's short deadlines (Figure 1.16).

FIGURE 1.16 Photo of a flight deck showing the large amounts of equipment needed for large projects.

One of the biggest benefits of UHPWJ is being able to accomplish blasting operations next to a zone that is receiving a coating (Figure 1.17).

FIGURE 1.17 Photo showing an area (left) that has just been blasted for primer application. The zone next to it (right) is being prepped for nonskid application.

1.4 Power Tool Cleaning

When there are areas inaccessible to UHPWJ or centrifugal blast equipment, power tool cleaning down to the bare metal is necessary; usually, these are deck edges or areas on the deck that do not receive nonskid systems but serve other purposes. CVNs have impact pads, arresting gear shives, and jet blast defectors (JBDs), which are located in open areas on the flight deck. These areas have power tool work accomplished around their perimeter.

When accomplishing power tool cleaning (Figure 1.18) to bare metal, it is important not to burnish the surface. Burnishing the surface will create a surface that is too slick, resulting in coatings failure. Failure would be the result of not having mechanical adhesion due to lack of profile.

Depending on the type of metal alloy that is being surface prepped, following are the power tools used:

- Handheld deck crawler—for steel only
- Manual walk-behind Descobrader[8]—for steel only
- Needle gun—for steel and aluminum
- Sander using #16 aluminum oxide grit disks—for steel and aluminum

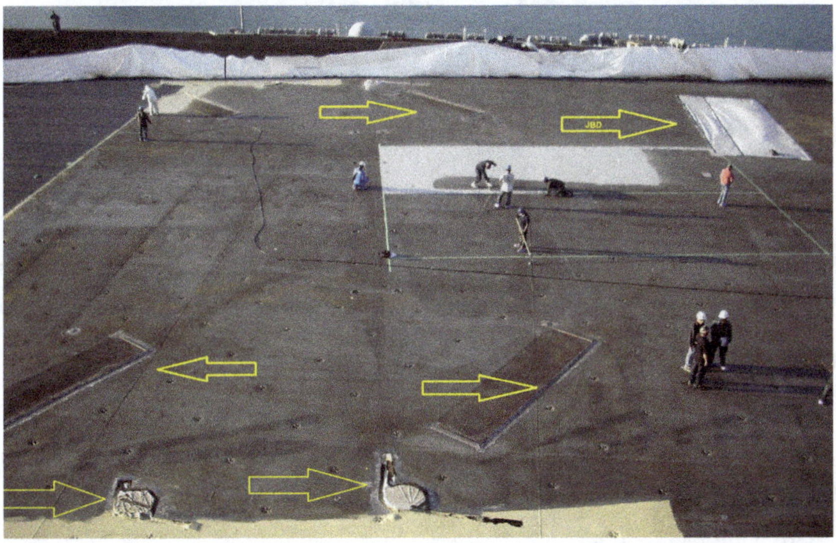

FIGURE 1.18 **Photo of CVN wire area; the yellow arrows show areas that receive power tool cleaning around their perimeter.**

Per NAVSEA Standard Item 009-32, the required standard to achieve when power tool cleaning to bare metal on military naval vessels is to meet SSPC-SP 11, "Power Tool Cleaning to Bare Metal."

1.4.1 SSPC-SP 11 Definition

- A bare metal power tool cleaned surface, when viewed without magnification, shall be free of all visible oil, grease, dust, dirt, mill scale, rust, coating, corrosion products, and other foreign matter, with the exception of trace amounts of coating and corrosion products in the lower portion of pits on pitted substrates (see Notes 8.1 and 8.3 in the standard).
- Acceptable variations in appearance that do not affect surface cleanliness as defined in Section 2.1 include variations caused by type of steel, original surface condition, thickness of the steel,

8. "Descobrader" is the industry term for the abrader originally made by Desco Manufacturing, Rancho Santa Margarita, CA.

weld metal, mill or fabrication marks, heat treating, heat-affected zones, or the texture/features associated with the use of a particular power tool.

- Deck edge power tooled cleaned to SSPC-SP 11 using a deck crawler and deck sander with a 16 grit disk, meeting the minimum-required 2.0 mil profile for a critical coated surface (Figure 1.19).

FIGURE 1.19 Photo of a deck edge power tool cleaned to SSPC-SP 11 using a deck crawler and deck sander.

1.4.2 SSPC-VIS 3

SSPC-VIS 3 is a visual comparater for power tool–cleaned surfaces, which is made up of photos for each standard. These photos have been taken of power tool-cleaned surfaces depicting four rust grades of carbon steel.

NOTE: The VIS 3 guideline photos were taken in a lab setting under optimum conditions.

The following are the standards visually represented in SSPC-VIS 3:

- SSPC-SP 2, "Hand Tool Cleaning"—until tightly adhered
- SSPC-SP 3, "Power Tool Cleaning"—until tightly adhered
- SSPC-SP 11, "Power Tool Cleaning to Bare Metal"—Paint and rust in pits only (Figure 1.20)
- SSPC-SP 15, "Commercial Grade Power Tool Cleaning"—33% staining allowed

Rust grades identified in SSPC-VIS 3:

- A—Steel surface completely covered with adherent mill scale; little or no rust visible
- B—Steel covered with both mill scale and rust
- C—Steel surface completely covered with rust; little or no pitting visible
- D—Steel surface completely covered with rust; pitting

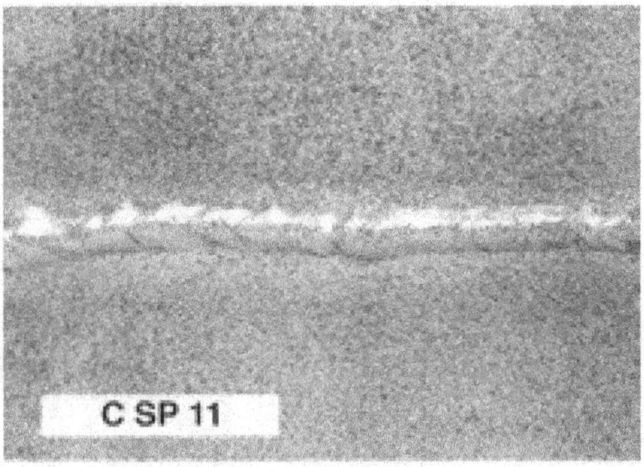

FIGURE 1.20 SSPC-SP 11 power tooled surface.

1.4.3 Contingency Paragraphs in Standards

NAVSEA has contingency paragraphs in SSPC Standard Item 009-32 for such scenarios if blast equipment cannot access the location or because the size or design makes it impossible to use non-skid blast equipment that would normally be used, i.e., a walk-behind centrifugal abrasive blaster or UHPWJ deck bug/mower.[9] Examples of ship locations that are inaccessible to nonskid blast equipment, and the reasons for the inaccessibility, include

- Ribbed ramps—design issue
- Interior passageways—location
- Mooring stations—location
- Mast decks—location and size

The NAVSEA surface preparation contingency in place for areas where deck blast equipment cannot access follows (quoted from the standard). NOTE: This only applies to areas that receive nonskid on U.S. Navy ships.

- For nonskid areas that abrasive blast equipment or waterjet equipment cannot access, substrate shall be prepared to SSPC-SP 11, except that minimum profile shall be 2 mil where accessible. The underside of tie-down bars [see Chapter 2, Figure 2.2 in this book] shall be prepared to SSPC-SP 2 or SSPC-SP 3, however, feathering requirements are waived, while the remainder of tie-down shall be prepared to SSPC-SP 15.

1.4.4 SSPC-SP 15 Definition

Below are the applicable paragraphs (numbered and quoted directly) from this standard, "Commercial Grade Power Tool Cleaning."

- 2.1—A commercial grade power tool-cleaned steel surface, when viewed without magnification, shall be free of all visible oil, grease, dirt, rust, coating, mill scale, corrosion products, and other foreign matter, except as noted in Section 2.2.

9. The "deck bug/mower" is either an electric or hydraulically run walk-behind piece of equipment used with the UHPWJ bar. It contains eight, high-pressure, specially made, cut-sapphire tips that shoot out water jet streams at 40,000 PSI, which removes the nonskid from the substrate or deck.

- 2.2—Random staining shall be limited to no more than 33 percent of each unit area of surface as defined in Section 2.3. Staining consists of light shadows, slight streaks, or minor discolorations caused by stains of rust, stains of mill scale, or stains of previously applied coating. Trace amounts of coating and corrosion products are permitted to remain in the lower portions of pits if the original surface is pitted [Figure 1.21 below].

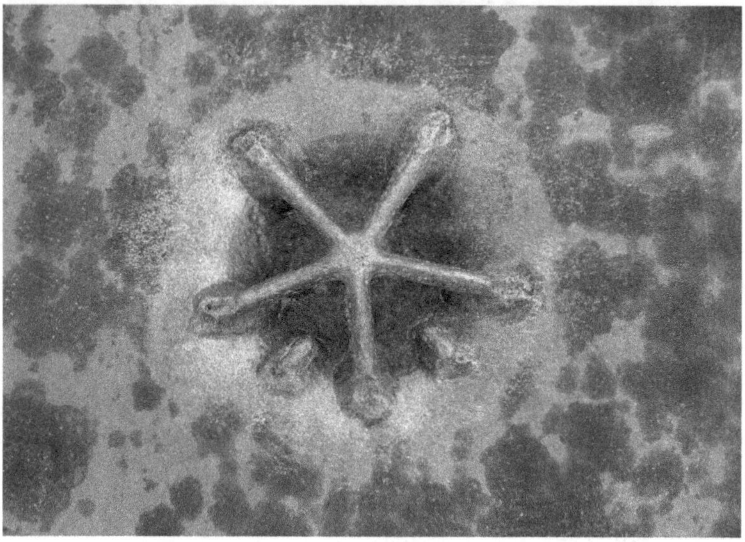

FIGURE 1.21 Photo of tie-down surface prepped to SSPC-SP 15. However, under the tie-down crossbars, it only has to meet the requirement of SSPC-SP 2.

2 Abrasive Blasting Different Alloys and Blast Media Selection for Nonskid Systems

2.1 Focus

This chapter is focused on abrasive blasting in relation to implementing the required profile. When accomplishing surface preparation on U.S. Navy vessels, many variables must be considered.

- Regarding the existing nonskid to be removed, is it tightly adhered and is it heavily applied?
- What is the substrate alloy?
- What is the high yield strength "hardness" of the alloy?
- Are the locations' structural design accessible to centrifugal blast equipment?

These questions are important, and knowing the answers can be the difference between damaging Navy property and/or doubling the work effort involved.

Knowing these factors can help mitigate

- Effects to the schedule
- Damage to the ship
- Use of the wrong blast media
- Producing improper profile depths
- Excessive personnel hours
- Costs

2.2 Blast Media

Choosing the correct size blast media is imperative for achieving the proper profile. Knowing the substrate's chemical makeup, i.e., the "alloy", is also imperative so that a blast media is not used that can cause galvanic corrosion because of dissimilar metals. The high-yield strength of the alloy is important to know as far as density and hardness are concerned. Knowing this, the correct size and chemical makeup of the blast media can be chosen, which will generate the desired or required profile depth.

When working for the U.S. Navy, contractors must ensure they use products that meet the Military Specification (MIL-SPEC). The best way to verify if the correct blast media is being used is to verify the requirements of NAVSEA Standard Item 009-32.

For the use of MIL-A-22262 abrasives, the contractor should ensure the blast media is on the Qualified Product Data List. This list can be found in a web search for the Naval Surface Treatment Center, https://www.nstcenter.biz.

The following requirements for blast media can be found in NAVSEA Standard Item 009-32 (quoted directly from the standard):

- Maintain a current copy of material certification of abrasive blast media conforming to MIL-A-22262, A-A-1722, or A-A-59316 for reference by the SUPERVISOR. Copy shall be available prior to blasting. MIL-A-22262 abrasives must be listed on the Qualified Products List (QPL), or the repair activity shall have written notification from NAVSEA indicating pending listing on the QPL. Submit one legible copy, in hard copy or approved transferrable media, to the SUPERVISOR upon request. For A-A-1722 or A-A-59316 abrasives, a complete data package demonstrating compliance with the requirements must be provided by the supplier to the procuring activity. Exceptions are listed in 009-32
- Recyclable Encapsulated Abrasive Media material conforming to SSPC-AB 4 may be used as an alternative to obtain SSPC-SP 10 or SSPC-SP 11 cleanliness.
- Recyclable ferrous metallic abrasive materials conforming to AB 3 of 2.5 (the standards for abrasives on navy ships) may be used as an abrasive blast media for steel substrates. Cleanliness of recyclable ferrous metallic abrasive materials shall be measured and maintained in accordance with the requirements of AB 2 of 2.5.
- For requirements specified in paragraph 3.1.3.3 of 009-32, maintain a current copy of the results of the quality control requirements of Paragraph 6 of AB 2 and quality assurance test required by Paragraph 5 of AB 3 of 2.5 for reference by the SUPERVISOR. Submit one legible copy, in hard copy or approved transferrable media, to the SUPERVISOR upon request.
- Steel shot, steel grit, and SSPC-AB 4 media containing steel abrasive SHALL NOT be used for surface preparation on the Amphibious Assault LHA, LHD, LPD, and LSD[1] Class ships. SSPC-AB 4 media that does not contain steel abrasive or MIL-A-22262 abrasive may be used on flight decks and other areas of LHA, LHD, LPD, and LSD Class ships.
- Understanding 009-32 references of AB-3 and AB-2 requirements for steel abrasives.

2.2.1 SSPC-AB 3 Covers Required Objective Quality Evidence (OQE) or Documentation

A certificate of analysis is one of the types of OQE that can be submitted; it breaks down the following subjects:

- Chemical composition
- Mesh size
- Specific gravity
- Hardness
- Durability
- Test Parameters

1. LHA: Landing Helicopter Assault, LHD: Landing Helicopter Dock, LPD: Landing Platform Dock, LSD: Landing Ship Dock

2.2.2 SSPC-AB 2 Covers Required Testing and Required Documentation

The following tests are required according to this standard:

- Sieve Test, ensures recycled abrasive meets the required mesh size after use
- ASTM-7393 indicates oil in abrasives test (jar test)
- ASTM-4940 is the test method for conductimetric analysis of water soluble ionic contaminants in abrasives. **Shall not exceed 1000 µS/cm (microseimens).**

Table 2.1 are steel shot mesh sizes and the SSPC-AB 2 sieve test breakdown chart. The mesh size of the steel shot media goes as follows: the lower the number the smaller the mesh size, and the higher the number the larger the mesh size.

TABLE 2.1 **Steel shot mesh sizes and SSPC-AB 2 sieve test breakdown chart from ASTM E11-22. Table reprinted courtesy of Mission Janitorial and Abrasive Supplies (now owned by Brady Plus, Law Vegas, NV).**

Screen Analysis

	Opening		Percent (%) by Weight										
SAE Size	mm	inches	S780	S660	S550	S460	S390	S330	S280	S230	S170	S110	S70
7	2.83	0.1110	All Pass										
8	2.38	0.0937		All Pass									
10	2.00	0.0787	85% Min		All Pass	All Pass							
12	1.68	0.0661	97% Min	85% Min		5% Max	All Pass						
14	1.41	0.0555		97% Min	85% Min		5% Max	All Pass					
16	1.18	0.0469			97% Min	85% Min		5% Max	All Pass				
18	1.00	0.0394				96% Min	85% Min		5% Max	All Pass			
20	0.84	0.0331					96% Min	85% Min		10% Max	All Pass		
25	0.71	0.0280						96% Min	85% Min		10% Max		
30	0.600	0.0232							96% Min	85% Min		All Pass	
35	0.500	0.0197								97% Min		10% Max	
40	0.42	0.0165									85% Min		All Pass
45	0.35	0.0138									97% Min		10% Max
50	0.300	0.0117										80% Min	
80	0.180	0.0070										90% Min	80% Min
120	0.12	0.0049											90% Min

**Screen Opening Sizes and Screen Numbers with Max and Min Cumulative Percentages Allowed on Corresponding Screens. ASTM E-11 and ISO 55 Test Sieves.*

2.3 Carbon Steel

2.3.1 Accomplishing Centrifugal Blasting on Aircraft Carrier's (CVN) Flight Decks

Usually, abrasive blasting is required wherever the arresting gear cables are located in the landing area because the arresting gear cables are dragged across the deck during arrests (traps) of the aircraft. The arresting gear cables basically slap the deck profile flat after so many arrests.

The abrasive blasting is limited to the areas where the arresting gear cables constantly hit the deck. The approximate square footage normally abrasive blasted to reimplement profile is 3000–4000 square feet in the landing area.

To determine the type and mesh size of the abrasive to be used, the chemical makeup of the substrate and its high yield-strength hardness—based on the Rockwell scale(2)—must be known.

Following are the CVN flight deck attributes for selecting the types and sizes of abrasive:

2. The Rockwell scale is a hardness scale based on indentation hardness of a material.

- Flight deck chemical makeup (alloy) carbon steel
- High yield designated as (HY-100)
- Tensile yield strength 100K PSI (690 MPa)
- Hardness (Rockwell) C-25

For nonskid systems in blasted areas, the acceptable surface profile criteria is

- Minimum surface profile is 3.0 mil
- Maximum surface profile is 6.0 mil

The condition of the existing nonskid must now be assessed. The condition will also affect the size of the blast media chosen and/or the steps a contractor will take to remove nonskid. By knowing these factors—plus the information on the substrate's chemical makeup—the correct blast media can be chosen.

Note that for carbon steel, the type of abrasive used is steel shot.

Existing nonskid may be heavy and tightly adhered. This will affect the production rate of removing the existing nonskid using the correct blast media to implement the required profile. Some contractors have the UHPWJ equipment to remove the existing nonskid first, then they do the abrasive blasting using the correct mesh size of blast media.

If the contractor does not have UHPWJ equipment to remove existing nonskid prior to shot blasting, following are steps that can be taken to meet the required profile:

- Abrasive (S-460) can remove nonskid and yield a profile range of 5.0–6.5 mil at a steady production rate of nonskid removal.
- For a high production rate of removing existing nonskid, a larger steel shot has to be used (S-550). This size abrasive will produce above 7.0 mil.
- Once nonskid removal is complete, purge blast equipment of the large blast media. It is important to purge the hopper of the S-550 blast media because it will produce an out-of-spec profile range above 7.0 mil.
- Then, fill blast equipment with a smaller size blast media to implement the required profile. Steel shot (S-390) will yield the optimum profile range of 4.0–6.0 mil.

Below is the guideline for achieving optimum profile range in nonskid systems when choosing the type and mesh size of blast media for different high yield (HY) alloys on different classes of ships.

HY-100 steel decks:

- Flight Decks CVNs, LHD, LHA
- Steel shot S-390 profile range 4.0–6.0 mil
- ES-1200 stainless steel shot range 3.0–5.0 mil

HY-80 steel decks:

- DDG—Flight deck and forecastle
- CG—Forecastle
- Steel shot S-280 profile range 4.0–6.0 mil

2.4 Aluminum Substrates

Choosing an abrasive for aluminum substrates is an entirely different situation that requires different processes and standards to be followed. Aluminum is a much softer alloy than carbon steel. The Navy has implemented the use of the Al aluminum alloy H-116-5000 series, usually the 5083 grade, for their decks. For example, the Cruiser, Guided Missile (CG) flight decks, CVN elevators, and LCS class ship decks are composed of the aluminum alloy Al 5456-H116-5086-H116. The standard that needs to be used for these aluminum decks is SSPC-SP 17, "Thorough Abrasive Blast Cleaning of Non-Ferrous Metals."

- Stainless steel shot mix using ES-300 and ES-450: the mix should be 70% ES-450 and 30% ES-300. Because the ES-300 is a very small stainless steel, it is common for the blast equipment to dump[3] with these small sizes of shot blast media. Two to three passes may have to be made to meet the profile range of 4.0–7.0 mil.
- The centrifugal blast equipment must move rapidly over the aluminum surface using only half the blast amperage for the blast media to achieve the desired profile range.
- If blast equipment moves too slowly but at a high blast amperage, the generated high profile will not meet the required profile depth.
- To remove the nonskid on aluminum surfaces, it is not recommended that centrifugal blast equipment be used because it will not be productive using the ES-300/ES-450 mix. The larger ES-1200 stainless steel shot (or bigger) would be needed because the contractor would be moving at a slower pace to remove the nonskid if the smaller ES-300/ES-400 mix was used. This, in turn, would produce a deeper profile that will have to be brought back within specification. It is recommended that ultra high-pressure water jetting (UHPWJ) is first used to remove the nonskid, then the ES-300/ES-450 mix can be used for centrifugal blasting.
- The average profile requirement for Navy ship decks is 3.0 minimum to 6.0 maximum. NAVSEA has implemented a different maximum profile of 7.0 mil for aircraft carrier elevators and CG flight decks because they are aluminum.

Some naval structures that require nonskid are composed of three different metals: carbon steel, aluminum, and stainless steel, which can be slippery. Destroyer Guided Missile (DDG) and CG class ships have forward and aft vertical launch systems (VLS) composed of these metals, and the nonskid provides safety for personnel movements while they work on these systems—whether in port or at sea.

Because of the configuration of the VLS (Figure 2.1) the surface preparation options are fewer when it comes to the type of equipment that can be used.

- Centrifugal blasting is not an option because the VLS is not a flat structure.
- Open-wand UHPWJ can be used, although this will not be able to meet the requirement of WJ-2L for light rust because a closed-loop vacuum system cannot be used. The structure is not flat, so this will then meet the criteria for WJ-2M (moderate) to a WJ-2H (heavy) rust on the parts of the VLS that are carbon steel.
- The most common surface preparation accomplished on a VLS is dry abrasive open-nozzle blasting in a 100% containment.
- Because the open-nozzle blasting uses three different alloys—carbon steel, aluminum, and stainless steel—the correct blast media must be used. The best blast media to use is garnet

3. Dumping is when a walk-behind closed-loop centrifugal blast machine releases all the stainless steel shot prematurely because of its small mesh size.

because it is a mineral abrasive that will not imbed or damage the aluminum; aluminum is the softest of the three metals.

- The specific mesh size for the garnet to meet the required profile depth per NAVSEA will be Military (MIL)-A-22262.
- Garnet mesh mix of 30-60; most manufacturers have this mix readily available.
- This will produce a profile from 3.0–5.0 mil, meeting the requirements for nonskid systems.

FIGURE 2.1 Photo of the aft VLS open-nozzle dry abrasive blasted surface.

Another nonskid surface that requires open-nozzle blasting is thermal spray nonskid (TSN) systems. This system only applies to LHD and LHA flight decks where the landing spots are located. TSN needs an aggressive anchor tooth (profile) to adhere, so steel shot centrifugal blasting is not allowed per NAVSEA message R-111724Z, because possible straggler steel shot might be on the deck during flight operations and may cause foreign object damage (FOD). Because of these reasons, the option chosen to implement the required profile is open-nozzle blasting using an angular blast media vs. a spherical steel blast media (Figure 2.2).

FIGURE 2.2 Photo of SSPC-SP 5 surface preparation prior to thermal spray application. Note the high-heat-resistant tie-down covers that prevent the thermal spray nonskid from filling in the tie downs.

Because of the requirements of NAVSEA Standard Item 009-124 for TSN systems, the following required blast media shall be used to meet the required profile depth:

- The LHA and LHD flight decks are HY-100 carbon steel.
- Because of the chemical makeup and density of the flight deck, NAVSEA implemented the use of a very hard blast media to ensure the required depth is met. NAVSEA chose aluminum oxide as the blast media for surface preparation of the TSN system.
- Aluminum oxide blast media is a very hard wear-resistant material, registering 9 on the Mohs hardness scale, and can be reused up to 15 times in a media reclamation system. Although this media can be recycled, it is not allowed per NAVSEA Standard Item 009-124. The blast media used must be virgin material for each surface preparation.
    - Traditional nonskid systems required range is 3.0–6.0 mil. For TSN, the required profile range is 4.0–10.0 mil.
    - The required mesh size and mix must conform to SSPC-AB 1 of 2.3 or MIL-A-A-59316, and be a blend of 50% 16-mesh grit aluminum oxide and 50% 24-mesh grit aluminum oxide. This mix will meet the required profile range of 4.0–10.0 mil.

3 Surface Cleanliness Inspection

3.1 Successful Nonskid Coatings

For nonskid coatings to be successful and meet their intended life service, the cleanliness requirements prior to applying the first coat of primer are probably one of the most important steps in the process. To ensure there is a contamination-free substrate, there are multiple steps and tests that can be done to ensure the cleanliness criteria are met. This chapter is a guide for these steps and tests.

One of the industry's most common statements is that 75% of all coatings fail because of improper surface preparation cleanliness. When working on nonskid decks, there are steps in the process to visually ensure the deck is free of dust, debris, oil, or hydraulic fluid drips that may have leaked from equipment used for surface preparation.

1. Identify any drips of oil or hydraulic fluid, mark them, and then, using approved solvent on a lint-free rag, dab and twist the area until clean of the contaminant. This is only for small drips. If any area exceeds 6 sq in, it should be high-pressured water cleaned using 5000 PSI at a minimum to flush out the contaminant. Preferably, use closed loop ultra high-pressure water jetting (UHPWJ) because this would be a more productive and efficient way to remove oil or hydraulic contaminants exceeding 6 sq in.
2. Blow the deck down using dry, clean air. A blotter test should be accomplished meeting ASTM D4285; this will ensure that the air is moister and oil free. The deck should be blown down in the same direction to a location on the deck where the heavy dust and debris can be swept and then bagged or vacuumed into a container.
3. Once the deck has been cleaned of any contaminants, as a final check, there is a nonskid contractor trade secret of using sticky rollers to pick up debris the blow-down did not get. A sticky roller consists of a stick with a roller frame and a 3/8" paint roller nap; using containment tape, wrap the tape sticky side out around the roller nap. Then, depending on the size or square footage of the deck, have one to four personnel uniformly walk the deck dabbing the deck when they find straggler debris. The adhesive will pick up the debris. Once the adhesive is covered with debris, wrap another round of containment tape sticky side out (see Figure 3.1) and continue if needed.

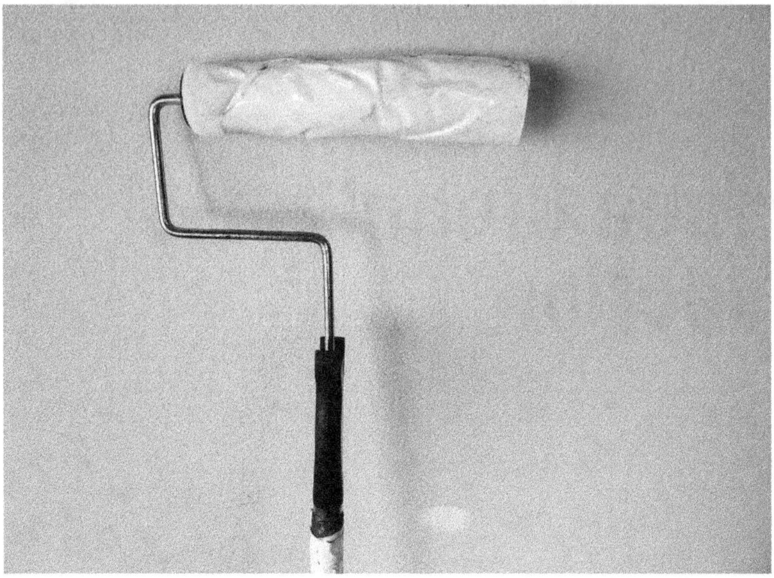

FIGURE 3.1 Photo of a sticky roller.

4. Accomplish a NAVSEA Standard Item 009-32 requirement for a visual assessment for dust in accordance with the ISO 8502-3 dust tape test. This is only required on flight decks and underwater hulls per NAVSEA Standard Item 009-32. The tape requirement is that it must be colorless, transparent, self-adhesive, pressure-sensitive tape with a width of 25 mm. The most common tape used is J-LAR[(1)] clear tape. The standard requires that the dust quantity rating be a maximum of 2, and the dust size also has a maximum allowance of 2. This requirement does not apply when using UHPWJ outlined in the SSPC-WJ 2L standard, unless 6 hours have passed since surface preparation has been achieved and no anticorrosive MIL-PRF-24667C primer has been applied. NOTE: When UHPWJ is used on flight decks and inaccessible areas are accomplished to SSPC-SP11, it is common practice that dust tape tests will be accomplished on those areas that have been power tool cleaned to bare metal.

Per the requirement of ISO-8502-3, to meet a dust size of 2, the particles shown on the dust tape test (Figure 3.2), should be just visible with normal or corrected vision (usually particles between 50 μm (micrometers) and 100 μm in diameter).

1. J-LAR is a registered trademark of Demco, Inc., Madison, WI.

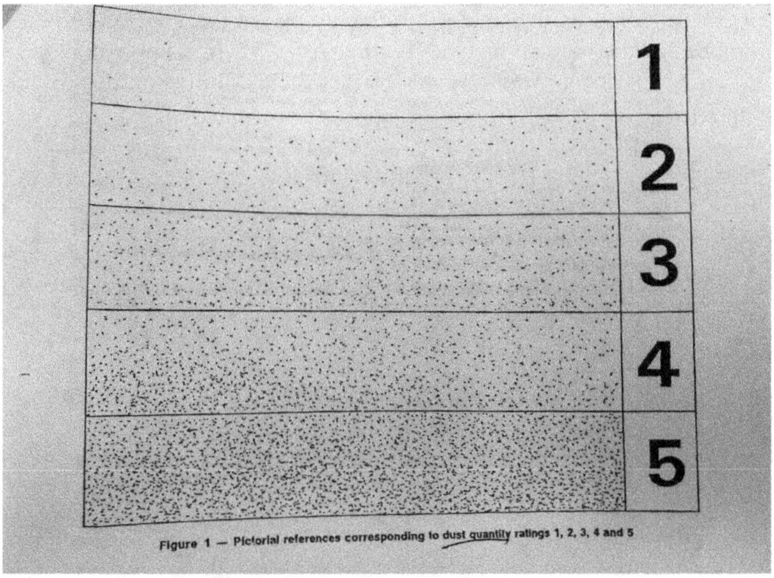

FIGURE 3.2 Photo of a dust quantity chart.

3.1.1 Testing for Nonvisual Contaminants on the Deck Substrate Surface

It is imperative that certain tests are accomplished prior to applying the first coat of anticorrosive primer. As stated in the opening of this chapter, some tests that will be discussed are required by NAVSEA Standard Item 009-32, and some tests are used in other industries outside of Navy ships—all of them, however, are relevant to nonvisual contaminants.

A required test by NAVSEA Standard Item 009-32 for nonskid systems is the Conductivity test (Figure 3.3), where conductivity is read as a number. The basis of the conductivity readings is as follows.

- The lower the conductivity reading, the lower the corrosion activity.
- The higher the conductivity reading, the higher the corrosion activity.
- The required maximum allowable conductivity reading is not to exceed 70 µS/cm (microsiemens per centimeter).
- If the reading is in chlorides, the allowable maximum reading is not to exceed 5 µg/cm^2 (micrograms/per square centimeter).
- If the conductivity of the deck plate is being measured, and only a conductivity meter is being used and the specification is requiring concentration in chloride, then the conductivity reading is converted to chloride using the following formula:

  E = 5 µg/cm^2 = 0.5, which is the maximum chloride allowable for nonskid

  S = Conductivity measurement reading

  V = Volume of water used to collect the sample; use 2 as the volume accounting for 1 mL of waste

  A = Unit area of the Bresle patch is 12.25 cm^2

Example: If the maximum allowable chloride reading is 5 cm^2, use 0.5 for E in this equation. This is then multiplied by the conductivity reading, e.g., 30 cm^2. This will be S in the equation, which should equal 15. Then, multiply this by 2, which will be V in the equation, which will equal 30. Then divide by 12.25 (A in the equation). This will equal 2.4 cm^2.

$E\,0.5 \times S\,30 = 15 \times V\,2 = 30/A\ 12.25 = 2.4\ \mu g/cm^2$

FIGURE 3.3 Photo of conductivity test on a Vertical Launch System (VLS).

Following are the ISO standards for the conductivity test.

- ISO-8502-6 is the extraction of soluble contaminants for analysis, known as the Bresle patch method.
- ISO-8502-9 is the field method for the conductometric determination of water-soluble salts.

3.1.2 Concentration of Chloride Testing

Chemical contaminants on a surface can include chloride, ferrous ions, sulfates, and nitrates. These chemicals or "salts" are deposited onto surfaces while the structure is in service. The atmospheric condition of Navy ships is salt water—one of the most corrosive environments to which an alloy can be exposed—hence, the Navy's requirement for testing conductivity. The maximum conductivity is 70 µs/cm (microsiemens per centimeter), which generally equates to 5 µg/cm^2 (micrograms per square centimeter)—the maximum allowable concentration of chloride per NAVSEA Standard Item 009-32. This ensures that the chlorides will not exceed a concentration of 5 µg/cm^2 by having a maximum conductivity of 70; this is their fail safe, therefore, conductivity is the main testing method on Navy ships, and along with the conductivity test method, is more user friendly than chloride test kits.

Following are chloride testing steps and test reading parameters:

- Latex Sleeves for Ion-specific Analysis—Chloride-Sulfate-Nitrate (CSN) Test Kit. If ion-specific testing for chloride, sulfate, and nitrate surface contamination is required, then a special kit is used to perform both the extraction and analysis. The CSN kit contains latex sleeves, premeasured extraction liquid, chloride-ion detection tubes (Kitagawa brand[2]), nitrate test strips, premeasured chemicals, and a meter for sulfate analysis.
- A premeasured amount of extraction solution is emptied into the latex sleeve and the sleeve is attached to the prepared, uncoated surface. An extraction is performed, the latex sleeve is removed, and the solution is analyzed for chloride nitrate and/or sulfate. For this method, the reading from the tube, strip, and meter is read in parts per million (PPM). This is not a requirement for nonskid systems, but it is a test method for concentration of chloride.

Following are the SSPC, ISO, and NACE[3] chloride testing standards:

- SSPC Guide 15, "Field Methods for Retrieval and Analysis of Soluble Salts on Steel and Other Nonporous Surfaces"
- SSPC Guide 24, "Soluble Salt Testing Frequency and Locations on New Steel Surfaces"
- ANSI/NACE SP0508, "Methods of Validating Equivalence to ISO 8502-9 on Measurement of the Levels of Soluble Salts"
- ISO 8502-9, "Preparation of Steel Substrates Before Application of Paints and Related Products—Tests for the Assessment of Surface Cleanliness—Part 9: Field method for the conductometric determination of water-soluble salts"
- ISO 8502-10, "Preparation of Steel Substrates Before Application of Paints and Related Products—Tests for the assessment of surface cleanliness—Part 10: Field method for the titrimetric determination of water-soluble chloride,"

There are other tests that can be accomplished to ensure cleanliness of nonskid coating systems. These tests are known as "water-break tests". These tests would be used in such instances as when nonskid coatings systems are contaminated prior to the next coat. Whether it is primer to nonskid, slip deck edges and tie downs prior to topcoat, or prior to color topping nonskid. These tests are not commonplace in the nonskid world, but they can be useful in case of contamination from hydrophobic materials, which are commonly used on Navy ships. These materials include hydrophobic films in the form of oils, hydraulic fluids, or any petroleum products that are difficult to see on existing coatings.

Following are water-break test standards (directly quoted from the standards):

- ASTM F21-20, "Standard Test Method for Hydrophobic Surface Films by Atomizer Test"–The atomizer tests are performed by subjecting the dry surface to be tested to a fine water spray. The interpretation of the test is based on the pattern of wetting. In the absence of hydrophobic films, the impinging water droplets will wet the surface and spread immediately to form a continuous water film. If areas of hydrophobic materials are present on the surface, the sprayed droplets will not wet the surface and spread, but will tend to remain as a fine droplet.
- ASTM F22-21 "Standard Test Method for Hydrophobic Surface Films by the Water-Break Test"– The water-break test is performed by withdrawing the surface to be tested, in a vertical position,

2. Kitagawa-America, LLC, Pompton Lakes, NJ
3. Although NACE and SSPC merged to form AMPP, "SSPC" and "NACE" appear here because the standards are still listed under these names.

from a container overflowing with water. The interpretation of the test is based on the pattern of wetting. In the absence of hydrophobic films, the draining water layer will remain as a film over the surface. In areas where hydrophobic materials are present on the surface, the draining water layer will break up into a discontinuous film within one minute.

When it comes to nonskid systems, it is of extreme importance that the cleanliness prior to each coat is inspected and required tests are accomplished. These basic steps and standards will help mitigate premature deck coating failures and help with the longevity of the nonskid system, which is the ultimate goal of all parties involved.

4 Nonskid Anticorrosive Primers and Compounds

4.1 What Are Nonskid Primers?

In this chapter, nonskid primers—also referred to as "barrier coats" or "anticorrosive primers"—will be discussed. The anticorrosive primer is the first step in the nonskid coatings system. Nonskid primers are epoxy-based organic coatings, a derivative from the epoxy chemical oxirane. The epoxy group is disubstitutable oxiranyl in which the two carbons are supposed to be connected with other carbon groups such as ketanes (C_2H_2O). Most nonskid primer epoxies are made from petroleum products, solvents, and pigment binders. Some contain zincs and other additives that provide its adhesion, durability, and abrasion resistance, which are needed for its use as a nonskid primer. This chapter will also cover polysulfide compounds used in nonskid systems.

4.2 Application Requirements

There are a number of application requirements that must be met for nonskid primers.

- NAVSEA has implemented requirements contingencies for the application of nonskid systems, most importantly "recoat windows", where the recoat window is the allotted time a coating can be recoated within a required time frame.
- Unlike most coatings applied on Navy ships, nonskid systems have the shortest recoat windows in the industry. Along with short recoat windows, most all deck work is the last thing to accomplish on all major ship repairs via PIA Dry Dockings, MAC-MO, PSAs, and SRAs.[1] This makes nonskid projects some of the toughest jobs when it comes to logistics and meeting scheduled deadlines.
- Depending on the nonskid manufacturer, some of the chemicals or additives may be different. This includes dry film thickness (DFT), which may vary, so DFT average requirements will not be covered. These requirements, however, can be found on the manufacturers' approved

1. PIA=Planned Incremental Availability, MAC-MO=Multiple Award Contract-Multi Order, PSA=Post Shakedown Availability, SRA=Ship Repair Availability

NAVSEA ASTM-F718 and on the National Surface Treatment Center website (https://www.NSTcenter.biz).
- NAVSEA has set recoat parameters for nonskid systems; each evolution is critical for the longevity of the coating's life cycle.
- The parameters start with the environmental conditions during the application process. As stated in NAVSEA Standard Item 009-32, the manufacturer has the requirements stated on ASTM-F718, but NAVSEA has more stringent requirements, so it is required that the most stringent requirements be followed:
    - Ambient temperature acceptable range is 55–100°F
    - Surface temperature acceptable range is 40–120°F
    - Deck temperature must be a minimum of 5°F above the dew point and rising.
        - Example: dew point 75°F, deck temperature 82°F
    - The maximum allowable humidity for nonskid systems is 85%
- The required intervals for taking the environmental readings shall be every hour during the application process. (Refer to NAVSEA Standard Item 009-32 for data logger requirements.)

It is important that the materials used are stored properly. These materials will meet the temperature parameters per NAVSEA and/or the manufacturer's requirements.

- Nonskid system materials must be stored in a cool, dry place, not exposed to freezing temperatures or direct sunlight, and must be stored in accordance with NAVSEA-reviewed ASTM F718s and/or the manufacturer's instructions.
- The minimum and maximum storage temperatures of the stored coatings should be documented over the 24 h period prior to use. Per NAVSEA Standard Item 009-32 it is required to document temperatures every 12 h. When taken manually, the average times documented are usually one at 0600 (6:00 am) and one at 1800 (6:00 pm) are. Using a data logger that can take minimum and maximum temperatures would be more advantageous for the deck plate Quality Assurance/Quality Control (QA/QC) Inspector.
    - The required storage range is 50–90°F.
    - For low temperature nonskid systems, both nonskid and primer, must be stored from 65–85°F.
    - The optimal storage temperature range is between 70–80°F.

4.3 The Application Process

Once the surface preparation standard has been established and the check point has passed, a mixing station is set up next to the paint pump. Setting up the mixing station is important—ensuring there is the correct amount of paint is key to a successful application.

4.3.1 Setting Up the Mixing Station

The first step in the application process is setting up the mixing station.

For most nonskid primers, the manufacturer's kits come in 5 gal containers. There are manufactures selling 4 gal kits or 3 gal kits, but for this chapter, 5 gal kits will be used in the examples. When it comes to primer application, knowing the square footage of the deck that will be primed is key. This will tell how many kits are needed to pull from the paint storage. For example, if there is a surface preparation area of 10,000 sqft, how many 5 gal kits will be needed?

The guideline for nonskid contractors is, on average, 1 gal of a nonskid primer will cover approximately 160 sqft with a DFT range of 4.0 mil. This is based on a range of a 75–85% solid coating. This is the safest way to ensure coverage.

NOTE: This is not a precise formula for all coatings. This is based on years of using and working with nonskid primers—it is a short cut that most nonskid contractors use.

Example:

- Use the square footage of 10,000 sqft.
- Divide by 160—this represents the 160 sqft per gallon that covers at 4.0 mil DFT. This equals 62.5 gal.
- Divide by 5—this represents the 5 gal kit—which equals 12.5. (If using a manufacturer that uses a 4 gal kit or 3 gal kit, then instead of 5, use 4 or 3 to divide by.)
- The number of 5 gal kits needed will be 12.5, which is rounded up to 13 kits.
- The guideline, however, is to always add two or three more kits, as a "cushion"—in addition to the number calculated—at the mixing station to account for any unforeseen issues.

4.3.2 Required Supplies

When ordering nonskid primer for a project, if the total square footage is known, the same basic formula as the one found in Section 4.2.1 can be used. Instead of using 160, however, use 140 sqft as the number to divide by. This will ensure enough product with a minimum cushion of 20%. As the percentage of solids go up, the percentage of cushion goes up also. Some contractors want to order "just enough," but this is not recommended. There are multiple scenarios that can take place on a project, and the contractor should want to always be in a position of adapting to these situations—not having enough product at the job site can be a logistical nightmare.

The correct size airless paint pump can be the difference between painting a 10,000 sqft deck in 1–2 h vs. 3–5 h. Nonskid primers percentage of solids are usually in the range of 70–90%. Knowing what type and composition of the required nonskid system is important; this will tell the percentage of solids of the specific primer being used.

The first step in the primer spray application is to ensure having the necessary cubic feet per minute (CFM) for the airless pump. If using an air compressor, the minimum recommended would be a 185 CFM compressor. This will keep the pump pressure between 90–120 CFM. The painter may ask for the pressure to go up or down during the application process, but it is normally within this range. The higher the percentage of solids of a coating, the thicker, i.e., more viscous (higher viscosity), the paint will be.

A paint pump is needed that can push the paint through the line without sputtering, or not having enough pounds per square inch (PSI) to push through the tip of the airless paint gun. It is important to use the correct size airless paint pump (Figure 4.1). Recommended airless paint pump size and spray tips for nonskid primers are the following:

- Minimum paint pump size, ratio of 74 to 1
- Paint tip size for decks larger and more-wide open than 1000 sqft, 631 or 635 tip sizes
- For smaller, more complex, decks with more interferences, 415–517 tip sizes

FIGURE 4.1 Photo of the mixing station and airless spray pump.

4.3.2.1 A Two-Part System

Nonskid primers are a two-part coatings system that cure by an exothermic chemical reaction. This means that after mixing the two parts together, the coating temperature will begin to rise (heat up), so it is important to apply the coating as soon as it is properly mixed. It is extremely important to keep the mixing station moving, i.e., mixing paint and continuing to change out each used paint can with a newly mixed can. The author has seen contractors in the past just leave the original mixed can that was first mixed remain in place with the pump suction tube in it, *as they continue to pour the newly mixed paint into this can*. This is a bad practice because the paint in that can is heating up through the exothermic chemical reaction and the paint may harden enough to cause clumps of curing paint that may get sucked into the pump suction tube and clog the paint tip. This will result in losing time to clear the gun. The can should be changed out if personnel find this to be the problem. Note, however, that while doing this, the paint is curing in the line as the chemical reaction heats up the coating, which means the line will need to be flushed out. The result will be losing time and paint because of something as simple as replacing each mixed can with a newly mixed can of primer to the paint pump suction tube.

4.3.3 Primer Application

The nonskid primers can be applied by spraying, rolling, and/or brushing. Usually, nonskid decks are sprayed, but there are exceptions with smaller decks that receive nonskid; sometimes it is not productive or convenient to set up a mixing station and spray pump when a deck only needs 3–8 gal of primer to meet the required number of mils. Most nonskid contractors recognize when it is more cost effective to brush and/or roll.

The decks identified below are usually applied via roller and brush because of size and interferences (Figure 4.2).

- Mooring stations
- 02 level fueling stations off the focsle
- Passageways and light traps
- DDG forward 02 level Nulka launcher deck
- Weapons magazine decks
- Rast machinery room decks
- Torpedo magazine decks
- Close-in weapon system (CWIS) decks
- Surface electronic warfare system (SLQ)-32 decks
- Bridge wing decks
- Mast decks

FIGURE 4.2 Photo of a small passageway where the contractor chose to apply the primer by roller and brush application. As shown, it would be cumbersome to set up a mixing station and spray pump for such a small area. There is also a risk of over-spray on the white bulkhead in such a small location. Roller and brush application was the correct course to follow.

4.4 Recoat Windows

As stated in the beginning of this chapter, nonskid systems have the shortest recoat windows of time in the naval marine industry. NAVSEA has implemented contingencies to aid in meeting recoat windows. One of these contingencies is the option to apply a stripe coat to welds and other areas—*prior to the full primer coat application*—to ensure coverage to areas that are either hard to reach or protrude from the deck. This is the only coatings system identified by NAVSEA that permits this action. The following are paragraphs from NAVSEA Standard Item 009-32 for the required recoat windows for nonskid primers:

- 009-32—Accomplish the requirements of 009-32 for stripe coat with the exception that stripe coat may precede prime coat.
- 009-32—For overcoating of stripe coat or stripe coating of the primer coat, refer to the applicable NAVSEA-reviewed ASTM F718. When the stripe coat is applied prior to overcoating with a full coat of primer, the stripe coat must be dry to touch in accordance with ASTM D1640, identified below.

ASTM D1640-14 Standard Test Methods for Drying, Curing, or Film Formation of Organic Coatings (used with permission from ASTM).

- Dry-To-Touch Time: Paragraph 7.5.1 of ASTM D1640, states "Continue testing after the set-to-touch time (reference paragraph 7.2 of ASTM D1640) has been observed. The film is considered dry-to-touch when it no longer adheres to the finger and does not rub up appreciably when the finger is lightly rubbed across the surface. Remove any coating from the fingertip immediately. The use of a glove, finger cots or the presence of freshly applied cosmetic products/ hand creams may interfere with the test results."

4.5 Stripe Coat Application

NAVSEA has given the option to apply the stripe coat and allow it to dry-to-touch prior to applying the full coat of primer. This saves a day in the recoat window process, and is advantageous in certain situations, usually during summer months and also in full enclosure or hangar bays. By accomplishing the stripe coat dry-to-touch, a day of coatings application or touch up is eliminated. This gives the contractor breathing room and saves an entire day (24 h) proceeding to the next step of nonskid application, keeping in mind that on critical coated surfaces, only three days (72 h) are allowed to cover the completed primer coat with nonskid. These are the advantages of the dry-to-touch stripe coat prior to the full coat of primer and how it can cut days down if a worker runs into the imminent danger of not meeting the period of performance deadline.

From NAVSEA Standard Item 009-32:

- 009-32—A second full coat of proprietary nonskid primer listed on the QPL for MIL-PRF-24667 may be applied if approved by the SUPERVISOR. The second full coat satisfies the stripe coat requirement.

FIGURE 4.3 Photo of stripe coat application prior to the full coat of primer on the USS *Carl Vinson* CVN-70.

4.5.1 Full Second Coat of Primer Satisfies the Stripe Coat Requirement

NAVSEA Standard Item 009-32 states that a full second coat of primer satisfies the stripe coat requirement. This generally is not used because it is not cost effective, especially on large decks. Nonskid contractors do, however, take advantage of this contingency when it comes to structures on certain classes of ships. The LPD and LSD older class ships have ribbed ramps. These ramps have 1 in high × 1 in thick steel brackets welded every 1 ft along the length of the ramp, port to starboard. A stripe coat of a contrasting color is applied to all edges or protrusions with a minimum of 1 in on each side of the edge or angle weld. In lieu of stripe coat, the contractor may apply a full second coat, which not only meets NAVSEA requirements, it also saves material and personnel hours. (A nonskid contractor can apply a full second coat in 45 min using an airless spray pump vs. 6–8 h of stripe coat application using brushes and rollers.)

- NOTE: Prior to applying a second full coat of primer in lieu of the stripe coat (Figure 4.4), an IDR/CFR[2] should be submitted requesting the supervisors' (Government representative) permission to invoke this NAVSEA Standard Item 009-32 contingency for a full second coat of primer to satisfy the stripe coat requirement.
- The NAVSEA Standard Item 009-32 nonskid application must begin within 36 h of completion of the final, full, primer coat application. For areas not listed in paragraph 3.7 of the standard, nonskid overcoating application must be in accordance with NAVSEA-reviewed ASTM F718. This statement is directed to noncritical decks. This is where the manufacturers' recoat parameters must be consulted. The NAVSEA standard has more stringent recoat windows for critical decks, which are identified in under paragraph 3.7. If the location receiving nonskid systems is not identified in paragraph 3.7, then the NAVSEA-reviewed ASTM-718 will be the recoat window directive because there are different manufacturers that produce nonskid coatings, and some of their recoat windows will vary. These processes are not covered in this book; know,

2. U.S. Federal Government independent dispute resolution/condition found report

however, that for every manufacturer's product being used on a noncritical deck, the NAVSEA-reviewed ASTM-718 of that specific manufacturer must be consulted.

FIGURE 4.4 Photo of a ribbed ramp second full primer coat application in lieu of a stripe coat, on the USS *Pearl Harbor* LSD-52. Photo courtesy of Jose M. Sanchez.

These are the areas that receive nonskid systems identified in NAVSEA Standard Item 009-32 under paragraph 3.7. Note that this standard does change annually, so to ensure compliance, the fiscal year (FY) that the specification requires must be checked. This book is using 009-32 FY23 for locations listed in paragraph 3.7:

- Flight Decks, this includes rast track plates when removed from the flight deck
- Hangar bay decks
- Vertical replenish decks (vert-Rep-decks) usually CG and DDG focsles and CG fantails
- CVN aircraft carrier elevators (ACE)-1, ACE-2, ACE-3, ACE-4
- DDG 05 level weather decks
- Vertical launch systems (VLS)

4.5.1.1 Solvent Wiping of Primer Coat

For critical surfaces, if the nonskid application begins within 36–72 h after completion of the final full primer coat application, the primer coat must be solvent wiped with the solvent required by the NAVSEA-reviewed ASTM F718.

Understanding why NAVSEA 009-32 requires the contractor to solvent wipe the primer after 36 h of cure time has passed prior to nonskid application. Understanding what the anticorrosive primer chemical makeup is will help make sense of these requirements. All nonskid primers are epoxies—organic polyamide thermoset coatings. Because the primer is a thermoset epoxy, the longer they sit and cure, the harder the coating gets. This means the chemical bond may be lost with the next coat (nonskid). The primer, however, is also a polyamide coating; when the solvent comes into contact with the primer prior to being fully cured, it will soften this coating. By solvent wiping the primer after 36 h has passed, the anticorrosive primer is softened, which ensures that the nonskid chemically bonds with the primer. Another reason for solvent wiping the primer is if it has been contaminated with dust or debris that is on top of the primer and not embedded; this will ensure cleanliness prior to the next coating (i.e., nonskid).

Per NAVSEA Standard Item 009-32, if nonskid application begins within three to seven days after completion of final full primer coat application, the primer coat must be solvent wiped with solvent required by the NAVSEA-reviewed ASTM F718. It is then lightly abraded, solvent wiped again, and a mist coat (1–2 mil) of primer must be applied and allowed to cure to recoat.

After 72 h (three days), NAVSEA Standard Item 009-32 requires the contractor to solvent wipe, lightly abrade, and apply a mist coat of primer 1–2 mil thick prior to the next coat. This is due to the fact that the primer has cured hard because it is a thermoset coating; the possibility of not having the chemical bond means that the nonskid will not adhere to the primer. Therefore, when the timeframe is lost for the chemical adhesion of the coatings, a mechanical adhesion for the mist coat must be created. This is the reason for the light abrading using 80 grit sandpaper. It is not advisable to use a larger grit sandpaper so as to not breach the coating to bare substrate. Once the coating has been abraded, creating an etched surface on the coating gives the mist coat an anchor to grab onto. The mist coat will then provide the chemical bond between the primer and the nonskid.

Per NAVSEA Standard Item 009-32, if the primer coat is not overcoated with nonskid within seven days of final full primer coat application, the primer must be removed, and the surface preparation repeated. For zone tie-in areas where the primer is to be overcoated with itself (up to 12 in of overlap), the recoat window must be in accordance with the NAVSEA-reviewed ASTM F718; the primer must be solvent wiped with the solvent required by the NAVSEA-reviewed ASTM F718, then lightly abraded, then solvent wiped again.

If the primer has sat for seven days because it is a thermoset coating, it is considered cured hard, and the window for using solvent and/or a mist coat using mechanical adhesion by abrading has been lost. Because flight decks are a critical part of any Navy ship, NAVSEA and the U.S. Navy will not take the chance of endangering ships' personnel or aircraft, so the area or zone must be resurface prepped in accordance with NAVSEA Standard Item 009-32 requirements. Per the standard, aircraft carrier landing areas not overcoated with nonskid within 72 h of primer application completion must have surface preparation repeated.

On CVN aircraft carriers (such as in Figures 4.5 and 4.6) where the wire area (arresting gear) is for aircraft landings or traps, the 72 h requirement is identified for this specific area beginning 1 ft aft of wire #1 to 300 ft forward starting from wire #4 within the port and starboard foul lines. This is the most critical part of an aircraft carrier flight deck, and the Navy will not risk the lives of pilots nor of losing aircraft. NAVSEA Standard Item 009-32 does not allow abrading and mist coat in the landing area; the standard considers the primer cured hard after 72 h, so the primer must be removed and the area resurface prepped in accordance with the requirements of the standard.

FIGURE 4.5 Primer application on the USS *Carl Vinson* CVN-70 wire area. Photo courtesy of Hugo Machuca.

FIGURE 4.6 Photo of USS *Carl Vinson* CVN-70 wire area in full coat of primer. Photo courtesy of Danny Barbosa.

4.5.2 What Can Affect the Primer?

When working with nonskid primers, there are times when weather or a Navy ship's design can affect the nonskid primer. Because most nonskid projects are accomplished in the open air, weather will play a factor during the curing stage. Some decks are above air conditioning units, or chill water piping may protrude through a deck—these would be part of the Navy ship design—but these factors can affect the primer, usually during its curing process. During the winter months after the primer is applied, the deck temperature drops and the dew point rises causing condensation on the primer. Usually, if the primer is applied early enough in the day, it gives the primer time to set up (exothermic reaction) before the temperatures drop at night.

The primer will usually show no signs of any issues, but if the primer is applied just before sunset and it is a cold night, the primer can be affected by these low temperatures causing condensation. This can affect the curing process—the reaction may appear on the primer and can manifest itself as gray cloudiness; gloss reduction; a greasy, waxy layer; or white crystals or patches. In some instances, it may not be visible, however, in contact with high condensation or water in cold temperatures, the coated areas can display opaque white marks. These marks can appear during or after curing. They are the results of two slightly different processes: blushing or blooming (see Section 4.4.2.1).

Naval ship design, as previously mentioned, can have air conditioning units on the bottom side of the deck or cold chill-water piping protruding through the deck. These produce cold temperatures and can also cause condensation on the deck during primer application and the curing process, which in turn causes the same blooming or blushing. Because of the ship's design, this issue can be mitigated by accomplishing a prejob walkthrough and noting this issue. This should be brought to the ship's force (i.e., personnel) attention, and they can possibly shut off the air conditioning unit during the nonskid project.

4.5.2.1 Blushing or Blooming of a Nonskid Primer

Two steps are taken when dealing with blushing or blooming of a nonskid primer:

- When blushing is present and has only affected the top surface of the primer. This means that the primer had time to set up and there are very light white marks, crystals, or patches on the deck, usually where the deck is warped or has a dip where moisture can puddle. When this is the case, solvent or water can be used to wipe the primer, and this usually will remove the white marks.
- When blooming occurs, this means the condensation or other moisture caused leaching of the primer compounds migrating them to the surface of the primer. This will result in a sticky surface that would not be easily removed with water or solvent. In this situation, use 80–100 grit sandpaper to sand the sticky compound off the primer. This can only be accomplished after the primer is cured hard enough to sand without damaging the primer. Once the sticky compound is removed, solvent wipe can be done.

The above bullets are best-case scenarios for blushing or blooming. Two things can compromise the primer's chemical makeup. The first is if nonskid primer was permeated by condensation or other moisture. When this happens, the entire deck turns a grayish white color, which does not come off with solvent or water. The second thing is if the sticky compound is excessive throughout the surface of the primer. These two things happen because the blushing or blooming is the effect of the curing agent or amine agent reacting to the condensation or other moisture in cold temperatures. The amine curing agent tends to absorb moisture in these conditions forming an ammonium carbamate. In the nonskid world, the basic terminology used is "amine blush/bloom" or "carbamation". If this is the case, then the primer will need to be removed in accordance with NAVSEA Standard Item 009-32 requirements.

FIGURE 4.7 (top and bottom) Photos of amine blushing where the primer was solvent wiped. The difference can be seen in the bottom photo where half was solvent wiped and the other half was not; the white was easily removed using solvent.

FIGURE 4.8 (top and bottom) Photos of amine bloom, not easily removed by solvent wipe. Abrading is needed using 80–100 grit sandpaper and solvent wipe.

4.5.3 Contaminants Affecting the Primer

Polyamide epoxies are generally durable and unaffected by weather and other contaminants when cured. It is during the application and curing stage, however, when the nonskid primers are usually at risk of failure or affected as stated previously with blushing and blooming when affected by weather and/or the ship's design. Contaminants that would affect the primer prior to and during application will be discussed in this section.

4.5.3.1 Tools for Inaccessible Areas

When accomplishing deck work, contractors use multiple tools for inaccessible areas like deck edges and tie-downs on flight decks and hangar bays. Most tools are electric and only need a source of power to operate. There are other tools used that are pneumatic, which run on air pressure, measured in cubic feet per minute (CFM). These tools are normally identified as chippers and needle guns.

Pneumatic tools need air tool oil; this is a synthetic product that lubricates the piston, which generates the impacting movement for the removal of coatings. It is imperative that the use of air tool oil be monitored on the amount used per needle gun or chipper; overuse of air tool oil will contaminate the substrate affecting the primer. The problem with air tool oil is it is hard to see on bare metal.

The worst area on which the author has seen this take place was a CVN flight deck where the contractor surface prepped 11,000 sqft of bare metal. This area had close to 600 tie-downs, and this was inside a tent. A guideline for nonskid contractors is once air tool oil has been introduced to the tool, run the tool for about 2–3 min over a cloth or rag letting it exude the extra air tool oil within the piston. Once it is observed there is no more air tool oil discharging from the tool, it is good to use to needle gun deck edges or tie-downs.

When the needle gun receives too much air tool oil, and a needle gun is being used for tie-downs on a flight deck, the needle guns will disperse drops of air tool oil around the tie-downs while they are being worked. The worker is unaware this is taking place and they move from tie-down to tie-down. This excess oil will affect the area of deck surface when primer is applied. When the primer is applied, and it comes into contact with the air tool oil, the coating will pull away from the oil leaving what looks like clear tiny bubbles. The industry terminology for this look is called "fish-eyes" (Figure 4.9). Because the primer was applied on top of the air tool oil, the coating must be removed. This will affect the contractor's primer inventory and schedule, so it is important that the contractor takes precautions to mitigate the amount of air tool oil used in pneumatic tools.

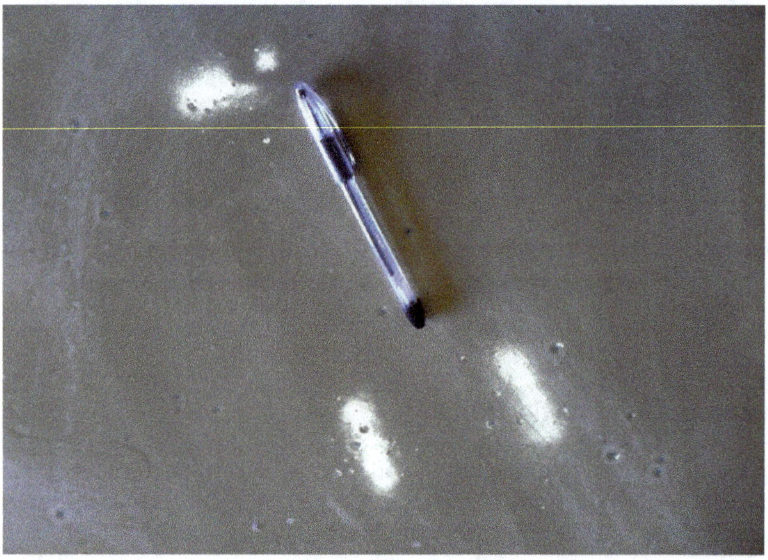

FIGURE 4.9 A closeup photo of the coating pulling away from the air tool oil leaving what looks like clear tiny bubbles known as "fish-eyes". The pen is used to get a visual of the size of the fish-eyes.

There are situations that can also contaminate the primer produced by the ship. Navy ships are always testing equipment or turning on the stacks that create smoke plums (i.e., "lighting off the stacks"). These are scheduled tests that most contractors are unaware of but can be detrimental to the surface preparation and application processes. When the Navy lights off the stacks, a plume of black smoke usually appears over the flight deck, and depending on the wind direction, this can and will affect any application of coatings. The black plume of smoke always carries soot from the stacks. Soot can best be described as a byproduct of the incomplete combustion of fueled carbon.

Multiple times in the nonskid industry, contractors have been caught off guard by the Navy lighting off the stacks during primer application or after the primer has been applied.

The specific curing stage of the primer is critical as to whether the soot can either embed itself into the primer or if it will only lie on top of the primer, which will result in minor stains that can be lightly abraded and solvent wiped. If the soot is embedded, then the primer will have to be completely removed. To mitigate this situation, it is important for the nonskid contractor to address the problem in the pre-job meeting with ship's force. The nonskid contractor should request a schedule of when they are going to light off the stacks. Usually, the Navy will not change their scheduled testing, but this will give the contractor notice on when *not* to apply any coatings. These situations can occur on any Navy ship but are most common on LHD and LHA class ships (Figures 4.10 and 4.11).

FIGURE 4.10 Photo of the ship's force lighting of the stacks, producing a plume of black smoke over the flight deck on the USS *Essex* LHD-2.

FIGURE 4.11 Magnified photo of soot in a tie-down.

4.5.4 Other Factors Affecting the Application Process

When applying the nonskid primer, there are many scenarios that can affect the application process. Nonskid contractors should always do their best to mitigate these when possible. These scenarios are the reason it is in the contractor's best interest to always order a minimum of 20% more product than what the square footage is for any project: if an issue arises where a nonskid contractor must remove a primer that has been newly applied and the manufacturers have long lead times for materials, the nonskid contractor might not be able to get the product to finish the project. This will affect schedule and worse yet, a Navy ship's scheduled deployment date.

4.5.4.1 Polysulfide Inhibitive Sealants

Polysulfide inhibitive sealants, or compounds, are used in nonskid systems when two dissimilar metals come in contact with each other. Their purpose is to prevent galvanic corrosion.

Polysulfide thixotropic Class B is a two-part, manganese dioxide–cured, polysulfide compound. The military performance number for this is 81733 (MIL-PRF-81733 Type-2); there are specific classes of ships and locations where polysulfide is used with nonskid systems:

- CVN aluminum aircraft carrier elevators (ACEs) with carbon steel tie-down cups
- CG class ships aluminum flight decks with carbon steel tie-down cups and vertical launch systems (VLS) composed of carbon steel, aluminum cell hatches, and stainless steel trunnions

- DDG VLS composed of carbon steel, aluminum cell hatches, and stainless steel trunnions

The polysulfide compound is applied at the intersecting seams of the VLS where the different metal alloys meet (Figures 4.12–4.15) to prevent galvanic corrosion. It prevents galvanic corrosion by eliminating moisture (rainwater, heavy dews, mists of salt water of the naval ships in the surrounding environment, etc.) from sitting on the cathodic alloy and anodic alloy that are in contact with each other. An example of this is the aluminum alloy deck surface on a CVN ACE with a carbon steel cup (Figures 4.14 and 4.15)—the aluminum elevator deck surface is the cathode, and the carbon steel cup is the anode.

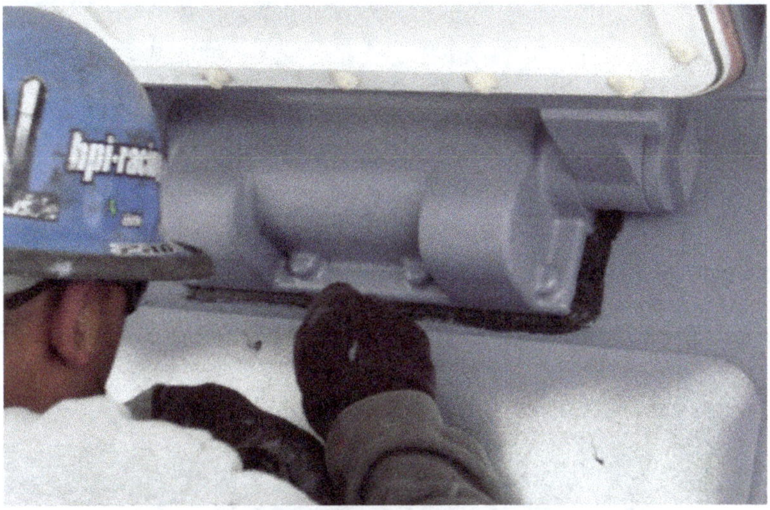

FIGURE 4.12 Photo of polysulfide applied to a trunnion on a VLS. The trunnion is stainless steel and the main VLS structure is carbon steel.

FIGURE 4.13 Photo of polysulfide applied to all VLS dissimilar alloys that come into contact with the carbon steel surface.

FIGURE 4.14 Photo of a CG aluminum flight deck with carbon steel cup, where the structural design was to keep the carbon steel cup and the aluminum flight deck separated. However, rainwater or any electrolyte would sit in the gap connecting the two different alloys and when the coating breaks down the galvanic corrosion will start. Photo courtesy of Felix Perez.

FIGURE 4.15 Photo of polysulfide applied between the aluminum surface and carbon steel cup. Photo courtesy of Felix Perez.

5 Nonskid Types and Compositions

5.1 Types and Uses

In this chapter, the different nonskid types and their uses for different locations on U.S. Navy ships will be reviewed. Each type of nonskid has been tested and designated by NAVSEA and the Naval Research Laboratory (NRL). The nonskid types and compositions are specified for unique areas or classes of ships. The requirements for this aggressive high-tech safety feature must meet the strictures in the NAVSEA Qualified Product Data List identified under Military Performance (MIL-PRF)-24667 and the NAVSEA-reviewed manufacturers' ASTM-F718. Most nonskids are epoxy-based coatings. The nonskid primers are *polyamide* epoxies that can be softened with solvent when not fully cured. Nonskid epoxies, however, are *polyamines*, which are more abrasion resistant and highly chemical resistant. Note that polyamine epoxies cannot be softened with solvents.

NAVSEA has also approved the polysiloxane nonskid system. Polysiloxanes are polymers with a silicon-oxygen backbone. They provide all the same attributes of the epoxy nonskid, but they also provide a high ultraviolet (UV) resistance leading to longer lasting color retention vs. their epoxy counterpart that becomes lighter (known as "chalking") over time.

There are several types of nonskid that are classified per NAVSEA and referenced in the Naval Ships Technical Manual (NSTM) 634 Rev-4:

- Type-1 High durability, rollable deck coating
- Type-2 Standard durability rollable deck coating
- Type-3 Standard durability, rollable resilient deck coating for use on exterior wooden decks, glass reinforced plastic (GRP), or aluminum decks where the deck flexes when it goes from hot to cold temperatures; Type-3 can also be used on warp decks because of the rubber membrane coating having self-leveling properties
- Type-4 Standard durability, sprayable, and rollable deck coating
- Type-5 Extended durability, rollable deck coating
- Type-6 High durability, fast cure, rollable deck coating
- Type-7 Fast cure, temporary repair, rollable deck coating
- Type-8 Low temperature cure, rollable deck coating

- Type-9 High temperature resistance deck coating
- Type-10 Submerged applications
- Type-11 Peel-and-stick applications

5.1.1 Type-1 and Type-2 for All Decks

Type-1 and type-2 nonskids are usually applied on all decks. They are considered the most common nonskids for general use and are polyamine epoxy-based coatings. Some type-1 and -2 nonskids may also contain Kevlar polymers that add additional abrasion and heat resistance. Type-1 nonskids have different compositions, which are described below.

NAVSEA has approved type-1 composition G (Comp-G) for general use on all decks. The aggregates are aluminum oxide, which are very hard and dense, and cannot be used in CVN landing areas.

NAVSEA has approved type-1 composition L (Comp-L) for limited use. This would be acceptable to use on an aircraft carrier (CVN) landing area. The most common aggregate in limited use nonskid is aluminum because it is a softer alloy and would not compromise the arresting gear cables in the landing area. The limited use nonskid is applied from 1 ft aft of wire #1 and 3 in outside of port and starboard foul lines and 300 ft forward of wire #4 (Figures 5.1 and 5.2).

NAVSEA has also approved type-1 and type-5, including type-7 as composition D (COMP-D) for low density or density controlled nonskid. The aggregate used is a thermoplastic compound and glass, which is approved for all L-class ships.

Type-1 and -2 nonskids have been designated with a coating life expectancy of 12 months per NAVSEA.

FIGURE 5.1 Photo of the aircraft arrest (or trap). An arrest or trap on an aircraft carrier is when the aircraft drops its tail hook (the black line descending from the aircraft's back) and attempts to land on the flight deck by trying to latch the tail hook onto one of the four wires (the four closely spaced, black horizontal lines) in the landing area on a CVN flight deck wire area. Photo courtesy of Lt. Commander David Martinez.

FIGURE 5.2 Photo of Comp-L and Comp-G where they meet 3 in from the foul line. The foul line markings always fall within the Comp-L nonskid. The general practice of most nonskid contractors is to apply the Comp-L nonskid 3 inches past the foul line where it meets the Comp-G nonskid.

5.1.2 Type-3 Flexible Nonskid

Type-3 nonskid is a flexible nonskid meant to be applied with a rubber membrane and is used on specific classes of ship where the ship's structure is not steel or aluminum and which could be damaged during nonskid removal—such as on wood or areas where cracking or fractures in traditional nonskid may occur because of flexing or heavy vibrations of the deck.

This type of nonskid is used on mine counter measure (MCM) wooden boats that have a GRP coating covering the wooden deck, followed by primer and a rubber membrane, followed by the compatible nonskid. This is for new construction of MCM class ships. Once the system has been implemented, the following process is used to remove and install the type-3 flexible nonskid system.

5.1.2.1 Removing Existing Flexible Nonskid

The process of removing existing flexible nonskid from MCM decks is to use a Desco[1] abrader to remove the nonskid down to the membrane. The objective is to lightly keep the abrader scaling the top of the nonskid without pushing down on the abrader as not to completely remove the membrane—this is to prevent damage to the GRP coatings. Once all nonskid has been removed to the membrane, the membrane will be uneven. The next step is to ensure all debris are clear from the membrane and then accomplish an SSPC SP-1 solvent wipe to ensure all dust has been removed. Then, apply a coat of nonskid primer followed by a skim coat of rubber membrane to fill in the uneven areas. Once this has been accomplished, the membrane will be a smooth and even surface, and ready for the compatible nonskid application (Figures 5.3 and 5.4).

1. Desco Industries, Chino, CA.

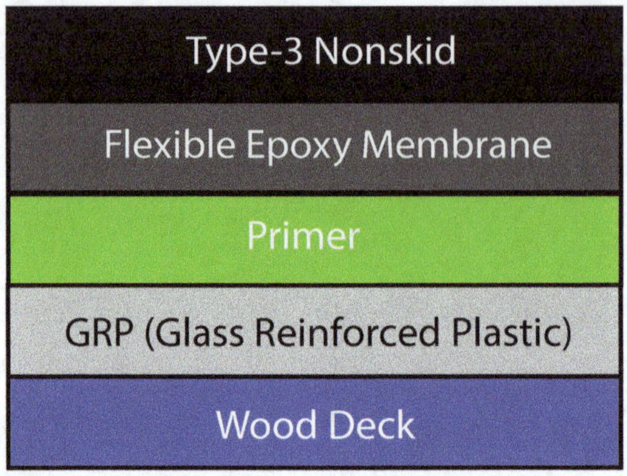

FIGURE 5.3 Colored breakdown of the type-3 nonskid system for MCM class ships. For new ship construction of MCM class ships, the nonskid contractor is only responsible for the primer, flexible epoxy membrane, and nonskid.

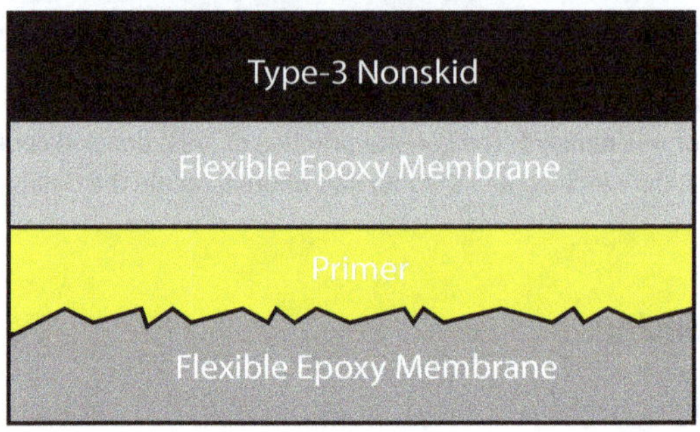

FIGURE 5.4 Colored breakdown of type-3 nonskid repairs on MCM class ships. The nonskid contractor only removes the nonskid to the membrane, applies a skim coat of primer, applies the flexible epoxy membrane, and fills in the jagged existing membrane leveling it out into a smooth surface for type-3 nonskid application.

There is another class of ship that has high vibration and an aluminum surface where traditional nonskid was cracking and delaminating. The entire craft of the ACU-5 Landing Craft Air Cushion (LCAC) is aluminum. This aluminum craft has two turbines at the aft end of the vessel that propels it forward (Figure 5.5). These turbines spin at such a fast rate that the entire aft section of the deck vibrates and would cause traditional nonskid to crack and fail over time. For these areas, a solution was developed (Bernard Maertz, Surface Technologies Corporation) of using a type-3 nonskid system that is mainly used on MCM craft. The new type-3 flexible membrane nonskid system was installed and proved that this system was able to withstand the vibrations during operations, which mitigated nonskid failure in this location of the craft (Figure 5.6).

The type-3 nonskid system has a required spread rate of 23–35 sqft per gallon.[(2)]

2. Although "sqft per gallon" is usually written as either "ft^2/gal" or "$ft^2\ gal^{-1}$", this discipline recognizes and uses this form, which is how it is used in this book.

FIGURE 5.5 Photo of turbines on an LCAC vessel. Photo courtesy of Hugo Machuca.

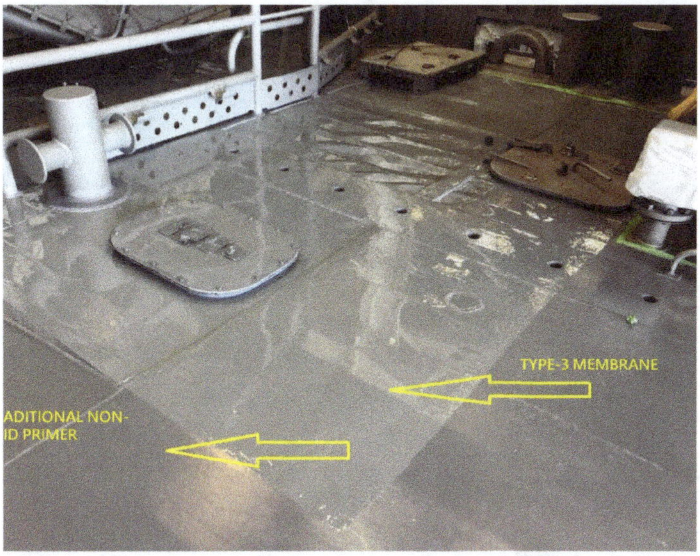

FIGURE 5.6 Photo of membrane applied in the specific area in front of the turbine just above the deck. This photo was taken prior to nonskid application. Unlike MCM class ships, which have to use the new type-3 flexible membrane nonskid system, the existing type-3 nonskid system on LCAC vessels is removed to the bare aluminum substrate for full application.

5.1.3 Type-4 Rollable Sprayable Nonskid

This system is most identified with spraying the nonskid vs. rolling the nonskid. Spraying nonskid comes with its complexities, one of which is using the right equipment, e.g., there is a Binks[3] closed style pot and a Pentek[4] open style pot. One of the main differences between rollable nonskid and sprayable nonskid is its spread rate. The rollable nonskid spread rate is 18–30 sqft per gallon vs. the sprayable nonskid range, which is 40 sqft per gallon.

Another factor of applying type-4 sprayable nonskid is the masking of bulkhead coamings (wheel stops) light fixtures. When accomplishing sprayable nonskid, more personnel hours must be factored in for masking and the materials needed to do so. Note that spray-on nonskid pots must have an agitator system in place to keep the grit constantly flowing to the gun as the spray-on nonskid is applied.

Traditional nonskid has to only mask off with 4 in tape and, when rolling the nonskid, the contractor only has to stay within the tape lined away from bulkhead coamings and fixtures (i.e., any protrusions from the deck).

Spray-on nonskid has a different texture vs. traditional roll-on nonskid. Spray-on nonskid appears more as an even, gritty, roughened surface resembling peel-and-stick nonskid vs. roll-on nonskid that has more profound peaks and valleys.

FIGURE 5.7 Photo of type-4 spray-on nonskid on an LCS flight deck. Notice there are no cross-rolled weld seams in the opposite direction (as can be seen with traditional rolled nonskid) because the spray-on nonskid is an evenly sprayed system over the entire deck. Photo courtesy of Felix Perez.

3. The Binks brand is owned by Carlisle Fluid Technologies, Scottsdale, AZ.
4. Pentek, Inc., Saddle River, NJ.

5.1.4 Type-5 Rollable Nonskid

There are two main compositions of type-5 rollable nonskid:

1. Type-5 Comp-G general nonskid can be applied on all decks, the same as type-1 and type-2 except for CVN carrier landing areas.
2. Type-5 Comp-L limited nonskid is used in landing areas as described in type-1 Comp-L.

NAVSEA specifications usually specify type-5 nonskid for critically coated surfaces. These surfaces are identified by NAVSEA in Standard Item 009-32 paragraph 3.7. The standard also specifies that the Government be present at all check-point and/or hold-point inspections.

One of the major differences between type-1 and type-5 nonskid is its service life. Type-1 is designated as a 12-month service life nonskid vs. type-5 nonskid has a NAVSEA designated service life of three years.

5.1.5 Type-6 and Type-7 Fast Cure Coatings

Type-6 and type-7—fast cure coatings—are not normally used by contractors, although the Navy usually has kits of types-6 and -7 onboard the ship for emergencies when there are nonskid repairs needed at sea. Because of their fast curing and durability, these aid the Navy in accomplishing patch repairs on decks such as flight decks. The fast-curing nonskid minimizes down time for flight deck operations.

5.1.6 Type-8 Low-Temperature Cure Rollable Nonskid

Type-8 low-temperature nonskid is used in cold weather. Contractors order type-8 nonskid when working in regions that usually have cold temperatures or if they are working on a project during the winter months. Type-8 nonskid is the only nonskid that can be applied with ambient temperatures as low as (minimum) 35°F and as high as (maximum) 90°F. With most traditional nonskid types, curing slows down and takes twice as long to cure when the ambient temperature drops below 50°F. Type-8 nonskid can cure within 12–24 h with ambient temperatures dropping to its minimum of 35°F. It is not recommended to apply type-8 nonskid below 35°F.

5.1.7 Type-9 High-Temperature Heat Resistant Nonskid

Type-9 nonskids are used in areas that can withstand direct heat for a certain amount of time that most traditional nonskids cannot:

- LHD LHA Spots 2, 4, 5, 6, 7, and 9
- DDG and CG harpoon launcher pads
- DDG and CG Nulka blast pads

LHD and LHA flight deck spots 2, 4, 5, 6, 7, and 9 take on immense heat during flight operations. The aircraft mostly used on these classes of ships is the Lockheed Martin[5] F-35B, which has a **vertical takeoff and landing**. Traditional nonskid was cracking during takeoff and landings because

5. Lockheed Martin, Bethesda, MD.

jet engines are facing down toward the deck on the designated spots. This also was a problem with the Boeing[6] V-22 Osprey, which also has a vertical takeoff and landing.

High-heat temperatures would buckle the deck, which cause the traditional nonskid to crack. Cracking nonskid that breaks off from the deck becomes what the Navy calls foreign object debris (FOD) and if pieces of failed nonskid get sucked into an intake on the aircraft it can cause engine failure. That may cause military personnel injury or even death—not to mention damaging a 35–40 million dollar aircraft.

The U.S. Navy, NAVSEA, and the Naval Research Laboratory (NRL) tested thermal spray coatings (Figure 5.8) as a way to mitigate nonskid cracking. Thermal spray nonskid has proven to work—it has minimal failures and can take the high-heat temperatures from the F-35B and V-22 Osprey.

FIGURE 5.8 Photo of applied thermal spray nonskid on the USS *Boxer* LHD-4.

The harpoon launcher pads and Nulka deck pads receive a rollable high-heat nonskid. The harpoon missile turbojet delivers 300 kg of thrust, putting out extreme high heat that traditional nonskid would not hold up to after so many launches. When launched, the Nulka decoy rocket also gives off a high-heat signature, and over time, traditional nonskid will break down.

The most commonly used type-9 on harpoon and Nulka launcher pads is FLEXFRAM 605NS[7] (Figure 5.9), where the "NS" indicates nonskid. The original formulation, FLEXFRAM 605, was developed for thermal protection of missile launcher systems from rocket engine blast, however, because of its toughness, erosion resistance, ablative efficiency, and adhesion, it should be considered for applications requiring short-term substrate protection from intense thermal and erosive environments. There is also another high-heat rollable nonskid identified as "Flamemaster".[8]

Flamemaster E340AF Blast Resistant Ablative Coating is used in areas subject to the extreme heat of jet-assisted takeoff. It is a room temperature, high-temperature resistant ablative coating. These

6. Bell Boeing, Amarillo, TX.
7. FLEXFRAM is a trademark of ACCRAbond, Olive Branch, MS.
8. Flamemaster is a registered trademark of Flamemaster Corp, Pacoima, CA.

high heat ablative materials are U.S. regulated and tracked products. When ordering this material, an ITAR[9] form must be filled out to identify on which U.S. Navy vessel it is to be used. Note that it is not authorized outside of the United States without U.S. Government authorization.

FIGURE 5.9 Photo of Nulka launcher pad FLEXFRAM 605NS application. Photo courtesy of Andre E. Lomeli.

5.1.8 Type-10 Rollable Nonskid

Type-10 nonskid is designated for immersion service and is used on submarine topside decks (Figure 5.10). As such, it is black to match the submarine color scheme. Type-10 nonskid meets all the same requirements of general nonskid with the exception that it has more rounded peaks for less drag on the submarine and the color is black vs. dark gray.

FIGURE 5.10 Photo of the topside nonskid on a submarine.

9. International Traffic in Arms Regulations (ITAR) (22 C.F.R. Parts 120-130), the ARMS Export Control Act (22 U.S.C. 2778).

5.1.9 Type-11 Peel-and-Stick Nonskid

Peel-and-stick coarse nonskid (known in the industry as just "peel-and-stick") consists of large abrasive particles in a tough durable resin bonded to a polymeric film. The reverse side is coated with a pressure sensitive adhesive covered by a removable protective liner. Peel-and-stick comes in sheets (see Figure 5.11) of different sizes. The most common sheet sizes used on U.S. Navy ships are 3 ft × 4 ft and 4 ft × 5 ft.

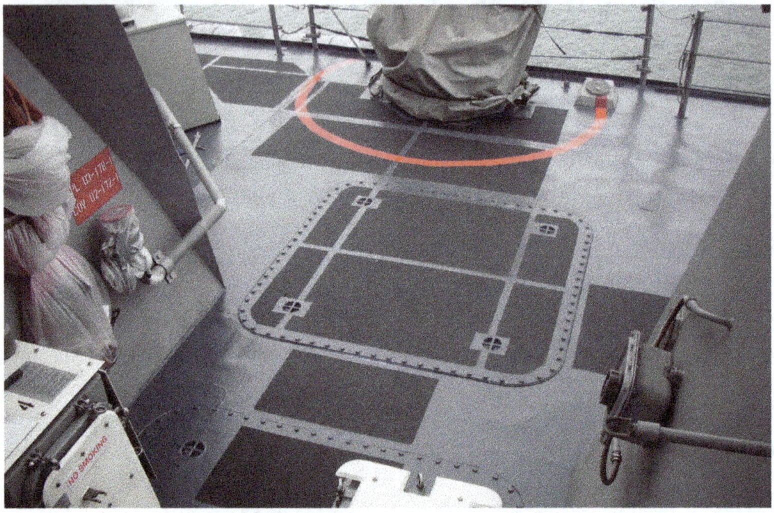

FIGURE 5.11 Photo of peel-and-stick installed.

Peel-and-stick falls under NAVSEA Standard Item 009-26 referencing attachment G in 009-26.

6 Nonskid Application

6.1 A Unique Type of Coating

Nonskid is a unique coating when compared to other coatings. Nonskid is a heavy, aggregated, polyamine epoxy that cannot be applied with traditional coating application methods, i.e., it cannot be applied using roller naps, brushes, or traditional airless spray guns. For the nonskid coating to produce a peak-to-valley antislip surface, a smooth, phenolic, hard-core, 9 in sleeve that slides on a 9 in roller frame must be used. This smooth phenolic hard-core sleeve is what produces the peek-to-valley surface. As the phenolic roller sleeve is pulled against the aggregated polyamine epoxy (nonskid), the aggregate is pulled into a peek and the epoxy resin lays flat creating the valleys. This action is what creates the antislip surface.

Nonskid recoat windows were covered in Chapter 4 because it is the nonskid primer that has to be coated within a specific time frame per NAVSEA requirements. In this chapter, environmental conditions during the application of nonskid, mixing requirements, application methods, and spread rates will be covered.

6.2 Environmental Conditions During Application

The manufacturer has the requirements for application stated on the ASTM-F718. However, NAVSEA Standard Item 009-32 has more stringent requirements—*and it is* ***these*** *requirements that must be followed, even if they differ from the manufacturer's.*

During nonskid application, the ambient temperature, deck temperature, and maximum allowable humidity are the same for both nonskid primer polyamide and nonskid polyamine:

- Ambient temperature acceptable range is 55–100°F
- Deck temperature must be a minimum of 5°F above the dew point and rising, e.g., dew-point

58°F, deck temperature 63°F

- Maximum allowable humidity for nonskid system application is 85%

However, these two epoxies differ slightly in their maximum acceptable surface temperatures where the primer polyamide epoxy can be slightly hotter:

- Surface temperature acceptable range is 50–120°F for the primer polyamide epoxy
- Surface temperature acceptable range is 50–110°F for the nonskid polyamine epoxy

The required intervals for taking the environmental readings (i.e., data logging) shall be every hour during the application process. Refer to NAVSEA Standard Item 009-32 for data logger requirements (covered in Chapter 4) because nonskid has the same parameters as the anticorrosive primers.

It is important that the nonskids used are stored properly. This is also covered in Chapter 4. Note that nonskid has the same parameters for storage as the anticorrosive primers:

- Required storage range is 50–90°F
- Low temperature nonskid systems (nonskid and primer) must be stored between 65–85°F
- Optimal storage temperature range is from 70–80°F

Once the primer and intermediate coat have passed their respective inspections (as well as any inspections required for the dry film thickness), the nonskid mixing and mixing station setup processes are the most critical aspects of nonskid application. If there are any problems, or one or two cans go out without the hardener (curing agent) to an area of 10,000 sqft or more, the problems will not be found until the nonskid is fully applied. Either two extra hardener bags will be found when a can count is done for spread rate, which will signify those two cans went out without hardener in them, or the extra hardener bags will not be noticed when the crew comes in the next day to "pull prep", meaning to pull up all the taped off borders for nonskid application once the nonskid is cured for "slip deck" application. "Slip deck" represents all the areas that were masked off, which did not receive nonskid. When looking at the newly applied nonskid, an area will be seen that is dull and does not have the same glossy sheen as the majority of the nonskid applied. This is a sure sign the nonskid cans in that area went out without the part B hardener.

6.3 Setting Up the Mixing Station

The first step in the nonskid application process is setting up the mixing station. There are major differences between the primer mixing station and a nonskid mixing station in the size, scope, and layout. Note that it can get chaotic if the crew doing the work, and the quality assurance and quality control (QA/QC) oversight on the mixing station, are both inexperienced.

1. Lay out plastic to protect the existing nonskid. Make sure to include a buffer of plastic from the line of cans at the front of the mixing station. The minimum buffer of plastic should be about 6 ft wide; this will be the traffic area where personnel get the freshly mixed cans and take them out to the nonskid applicators. It is important to protect this buffer area because it usually gets the messiest.
2. *This step is crucial*: laying out the 5 gal kits of nonskid. It is imperative that this be organized and set up comfortably enough for the nonskid mixers so they can move can-to-can without getting

tangled up and making a mess of the station (Figure 6.1).

- Example: A previously primed 10,000 sqft area is receiving nonskid. This will be a pretty good size mixing station. The 5 gal kits of nonskid should be laid out in rows of 10; each can should be a minimum of 1.5 ft apart in each direction side-by-side and behind as the kits are laid out in rows of 10.

3. Then, the team will start removing the metal can lids and pulling out the 1 gal hardener bags or cans, depending on the manufacturer being used. These hardener bags or cans should be laid out directly in front of each nonskid can. This will help keep track of the part B (hardeners) (Figure 6.2).

FIGURE 6.1 Photo of the nonskid mixing station setup with the cans evenly set apart and organized.

FIGURE 6.2 Photo of part B hardener bags directly in front of each can for tracking and ensuring each can receives a hardener.

6.4 Mixing Paddles

The proper mixing paddles must be used for nonskid materials. Nonskid mixing paddles must be double bladed (Figure 6.3). The double-bladed paddle creates a vortex (Figure 6.4) that ensures a proper mix of the viscous aggregated nonskid material. Normal mixing paint paddles do not create the same vortex that the double-bladed paddle can provide.

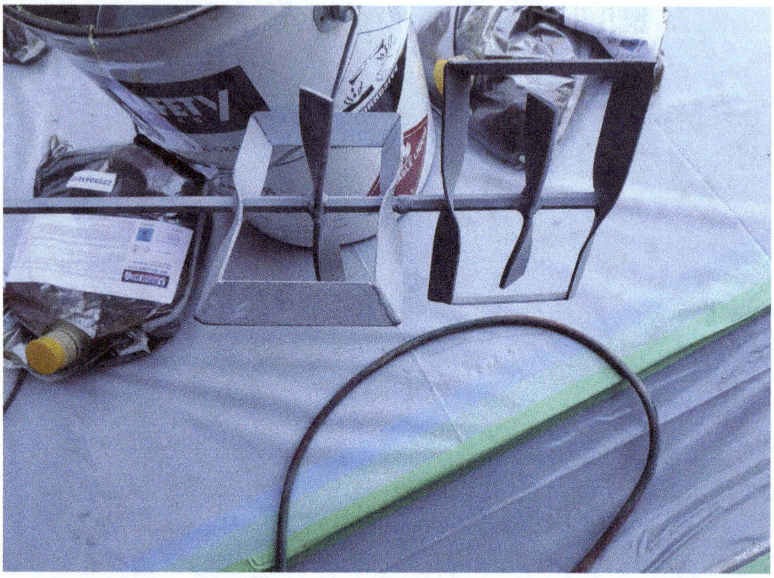

FIGURE 6.3 Photo of double-bladed paddles. Most contractors have them made by either welding companies or in-house welders. The paddles are always welded facing opposite directions, which intensifies the vortex during mixing.

FIGURE 6.4 Photo of a premixed can where the vortex—created while mixing with the double-bladed nonskid paddle—can be seen.

If a traditional paint paddle has to be used to mix nonskid, the mixing time should be extended longer than the ASTM-718 required mixing times to ensure the nonskid is properly mixed. Using a regular mixing paint paddle might work for a small patch of nonskid or a deck less than 300 sqft. On any areas larger than this, however, it is not recommended because it will affect the nonskid application and lengthen the amount of time needed to apply the nonskid. It also may cause "dry roll", which is when the mixed nonskid "pot life" (i.e., how long it is viable in the can) is nearing its limit; it then becomes too viscous and loses its fluidity, or "popcorning" occurs. Popcorning is the same as dry rolling with the exception of its appearance—literally popcorn-like. It is caused by the applicators over-rolling existing (i.e., already applied, drying, and curing) nonskid while applying new (fresh) nonskid. When over-rolling the old nonskid, trying to blend the curing nonskid with the new nonskid, the curing nonskid takes on the characteristic popcorn appearance; it loses its evenly dispersed look and resembles popcorn balls. It is best to overlap the new nonskid with the curing nonskid to blend them; this usually happens when doing weld-cross rolling.

Note that there is a NAVSEA Standard Item 009-32 requirement for the condition of the mixing paddle:

> The contractor shall visually verify that the nonskid mixing blade is free of previously cured coatings, contaminants, or rust prior to mixing.

6.5 Mixing Time Requirements

NAVSEA Standard Item 009-32 will refer to the manufacturer's ASTM-F718 for mixing time requirements. Once the nonskid mixing station is set up and cans are uniformly laid out, the lids are removed, and the part B hardener (Figure 6.5) is in front of each can. Because "10,000 sqft" is being used as the example deck's square footage, the best management practice for most nonskid contractors is to premix the first three rows. For nonskid application to a 10,000 sqft deck, it is best to at least have a minimum of three mixes going at once—more would be better, but no less than three.

FIGURE 6.5 Photo of part B hardener being poured into the part A nonskid can.

A nonskid mixing station needs to be orchestrated efficiently or things can get out of control quickly when dealing with this much product and movement in such a tight area. It is common practice to premix three rows of 10 before any part B hardener is dropped into the nonskid cans. The premixing times will be stated in the NAVSEA-reviewed ASTM-718. Most manufacturers set their premix time to 2 min per can. It is the contractor's responsibility to ensure they follow the manufacturer's instructions on the mixing times.

- Once 30 cans have been premixed, the hardener-dropper personnel can start putting the part B hardener into the first row of 10 cans. Once they have completed this, and 10 cans have the part B hardener in them, then two mixer personnel can be sent back to the first row of cans on each opposite side to start mixing the part B hardener and the part A nonskid resin in accordance with the manufacturer's time requirements. Most manufacturers set their full mix times between 3–4 min. Note again that it is the responsibility of the contractor to follow the mixing times of the manufacturer.
- Once the two mixer personnel are mixing the part A and part B nonskid materials, the hardener-dropper personnel (Figure 6.5) will continue to input the hardener following in step directly behind the premixer personnel; they continue the premixing operations while the other two assigned mixing personnel mix parts A and B together. It is *imperative* that there is an assigned person timing the premix and full mix of parts A and B, and it is *essential* to keep track of all movements on the nonskid mixing station.
- After parts A and B are mixed in accordance with the required mix times, each fully mixed can should be removed from the row of nonskid, putting it outside the rows for the people transporting the cans to the nonskid applicator personnel (a.k.a. "the applicators"). This is necessary because it is easy for a premixed can to get confused with a fully mixed can. This is where mistakes can take place if personnel lose track of what is taking place on the mixing station.
- Once the nonskid mixing station is in process, because 10,000 sqft is being used as the deck square footage, the best management practice for a deck this size is to use between 7–10 applicators. For this number of applicators, 10–15 cans should be mixed and sent out. This will give the mixing station time to keep up with the applicators.
- If only the first fully mixed can is sent out, then only one applicator will be applying while the other applicators wait. The one can of applied nonskid will start the curing process by exothermic reaction. If blending or seaming together new nonskid (known as "butting up") is attempted with the nonskid that has been sitting on a deck for more than 5–10 min, it will dry roll and cause the popcorn effect.
- It is imperative to put out enough fully mixed cans to line up the nonskid applicators so there is an evenly rolled application process while the cans continue to come out keeping the nonskid application process moving

6.6 Applying the Nonskid

The nonskid is applied using a smooth hard-core phenolic sleeve (Figure 6.6). When preparing for nonskid application, the applicator must ensure that the sleeve has no cuts or delaminating parts. If the sleeve is damaged in any way, this will cause skips or flat areas in the appearance of the applied nonskid. *Prior to any nonskid application, this should be checked and verified prior to use.*

Note that smooth hard-core phenolic sleeves only come in 9 in lengths. In tighter areas for nonskid application, contractors have cut the sleeve down to 4 in for use on a 4 in roller frame.

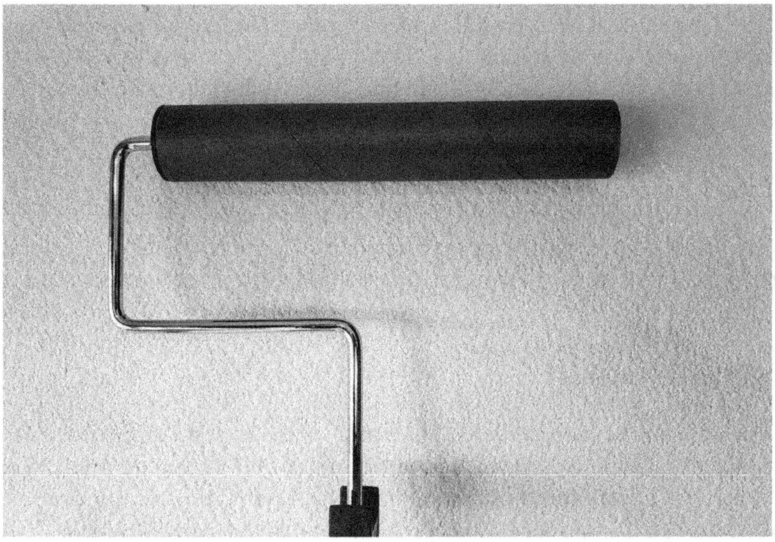

FIGURE 6.6 Photo of smooth hard-core phenolic sleeve on a 9 in roller frame.

NOTE: Knowing your required nonskid spread rate for nonskid application to a 10,000 sqft area will ensure there is enough product on the mixing station and the job site. NAVSEA Standard Item 009-32 stipulates the required spread rate range: for "general" and "limited" nonskid, it is from 18–30 sqft per gallon. On average, general and limited nonskid applies at a 25 sqft spread rate.

Most nonskid kits (regardless of manufacturer) come in 5 gal units. When it comes to nonskid application, knowing the square footage of the deck is key. This number will enable the contractor to calculate how many kits have to be retrieved from the temperature-controlled storage. Note, however, that dry film thickness cannot be done on nonskid because of its jagged, roughened surface and thickness. Because of this, the "paint formula" for nonskid cannot be used as it would for actual paint. The way to ensure enough nonskid kits are available for any project and mixing station is by the spread rate. This is how a contractor will know if the coverage needed will be obtained.

"Nonskid spread rate" is knowing how much deck needs to be covered with 1 gal of nonskid. Most nonskid contractors order at a 23 sqft per gallon spread rate. For nonskid applied during normal temperatures with a deck temperature range of 70–100°F, the nonskid spread rate, on average, normally ranges from 24–27 sqft.

Example: For 10,000 sqft, how many 5 gal kits of nonskid will be needed?

- Square footage of 10,000 sqft is determined.
- Knowing the nonskid kit is 5 gal, use the 23 spread rate as the starting number, then multiply by 5 representing the 5 gal nonskid kit. This equals 115 gal.
- The square footage of 10,000 sqft is divided by 115 (representing a 5 gal kit [previous bullet]). This equals 86.9 kits. Round this up to 87 kits.
- Then take the 87 kits and multiply by 5. The 5 represents the gallons per kit. This equals 435 gal of nonskid.
- Use the 10,000 sqft divided by 435 gal; this equals a 22.9 sqft per gallon spread rate.

- The 87 kits will give a 22.9 sqft per gallon spread rate (rounded up to 23), so this is the number of cans wanted on the mixing station.
- The 23 spread rate is mathematically a safe number; it will yield more than enough of what is needed knowing that a 25-spread rate will be obtained in actual practice, on average, which is common with general or limited-use nonskid.

When working in colder temperatures when the deck temperature is 50–65°F, less spread rate will be obtained, i.e., when applying on colder decks and ambient conditions. When ordering nonskid for these conditions, order at a rate of 20 sqft per gallon spread using the same formula.

6.6.1 Applying Type-3 Nonskid

For type-3 nonskid, the application method is the same as the general and limited nonskid with the exception of spread rate. Type-3 NAVSEA Standard Item 009-32 spread rate requirements are a minimum of 23 sqft per gallon and a maximum of 35 sqft per gallon. It is recommended to order the type-3 nonskid using a 25 sqft per gallon spread rate with the same formula.

When applying nonskid, NAVSEA has requirements when it comes to nonskid borders that receive slip deck coatings. NAVSEA Standard Item 009-32 states that nonskid must be applied to within 2–3 in of deck fittings and protrusions. Nonskid must be applied to within 4–6 in from coamings and deck edges, areas underneath protrusions (such as stowage racks or fire stations that are normally found on hangar decks), or coaming areas of decks not subject to vehicle or aircraft movement, need not comply with the 4–6 in rule for applying nonskid near deck edges.

The first step prior to applying the nonskid is to mask off the slip deck areas where there is no foot or vehicle traffic. For tie downs prior to nonskid application, it is not recommended to try to mask off using 2 in masking tape. This will not look uniform and will not be cost effective.

There are companies that sell precut adhesive circles that can be used: 6 in covers for countermeasure wash downs, 9 in covers for cross bar tie downs, and 12 in and 18 in covers for clover leaf tie downs, depending on the size of the clover leaf tie downs (Figures 6.7 and 6.8).

FIGURE 6.7 Photo of rows of cross bar tie downs covered with 9 in circles with an adhesive backing.

FIGURE 6.8 Photo of clover leaf tie down with the 18 in adhesive circle preparation pulled, leaving a perfectly round 2 in border around the weld seam. Photo courtesy of Jose M. Sanchez.

When applying nonskid, there are techniques to ensure the proper coverage is obtained and to ensure the nonskid is not being applied too thick or too thin. When applying nonskid by roller, the contractor must ensure enough pressure is applied to the roller's 6 ft extension stick for an even coverage of nonskid. This will mitigate leaving excessive nonskid on the deck. This also applies to not applying to much pressure and/or over-rolling the nonskid because this will thin it out, lowering the peaks giving a flatter appearance. Over-rolling the nonskid on deck with temperatures above 85°F will also start to dry roll giving the popcorn effect.

When the nonskid application is in full operational mode, there are nonskid best management practices that should be followed:

- When there are 10 people (i.e., applicators) applying nonskid, all applicators should be in their own lane, with "lane" meaning from weld seam to weld seam.
- The applicators must ensure the welds are farther ahead of the general application (Figure 6.9). This is because the weld seams must be cross-rolled in the opposite direction of the general application.
- The purpose of crossing the weld seams is to have the nonskid completely cover the weld so it does not leave a gap or so the nonskid is not applied too thick, which could happen if the nonskid is rolled in the same direction as the weld seam (Figures 6.10–6.11).
- NAVSEA Standard Item 009-32 requires that the weld is cross-rolled with a minimum of 3 in, to a maximum of 6 in, on each side of the weld (Figure 6.12).

FIGURE 6.9 Photo of nonskid application with the weld seams farther ahead of the general application. Photo courtesy of Hugo Machuca.

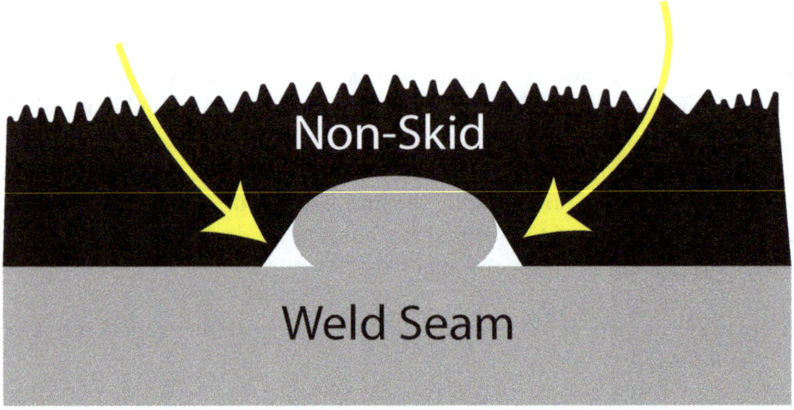

FIGURE 6.10 This graphic shows how if the nonskid is rolled in the same direction as the weld seam, it can leave a gap between the nonskid and deck (indicated by the yellow arrows pointing to the white triangles).

FIGURE 6.11 This graphic shows that cross-rolling the weld seam ensures even, tightly adhered coverage of the nonskid and the deck.

FIGURE 6.12 Photo of weld seam cross-rolled meeting the minimum of 3 in and maximum of 6 in on each side of the weld. The applicator applied 4 in on each side of the weld with uniformity.

6.6.2 Applying Nonskid to Ramps That Are Not Ribbed

Just as weld seams are required to get cross-rolled port to starboard, so do ramps. These types of ramps are usually on LHD, LHA, LPD, LSD, and LCS class ships and are applied port to starboard (left to right of the ship). This means that each entrance and exit of the ramp represents the forward and aft (front and back) of the ship. The reason for applying the nonskid port to starboard is these ramps are normally at a 45° angle whether it is being applied going up or down the ramp. When applying nonskid in any general direction, peaks and valleys flowing in one direction will be seen. As can be seen when cross-rolling the welds, the peaks to valleys flow in the opposite of the general direction of the nonskid application. This has been general practice for nonskid contractors for the past 30 years, and NAVSEA Standard Item 009-32 has now included this as a requirement.

Applying nonskid port to starboard on L-class ship ramps, creates opposite tread traction because forklifts or other vehicle tire treads will run forward and aft of the vehicle, while the nonskid on the ramp is rolled port to starboard. This will create opposite tread traction providing a more antislip surface for heavy vehicle traffic on the ramp. (Figure 6.13).

FIGURE 6.13 Photo of a forklift going up a ramp on the USS *Boxer* LHD-4.

6.6.3 Applying Nonskid to Rast Track Plates

There are other unique surfaces on naval ships, which receive nonskid that are not necessarily applied on the ship. There are rast track plates (Figure 6.14), which are part of a flight deck system. This system is designed to pull helicopters into the hanger bay once it has landed. The rast track plates play two roles: they are part of the flight deck, and a part of the rast track pully system. These steel plates are removable, so the nonskid work is normally accomplished off the ship (known as "off ship") when the rast track trough is receiving structural repairs and preservation.

FIGURE 6.14 Photo of rast track plates masked off for nonskid "off ship".

The rast track plates are general to two different classes of ships:

- CG class ship, which is an aluminum alloy flight deck
- DDG, which is a carbon steel flight deck.

Applying nonskid to the rast track plates (Figure 6.15) can be an arduous task of ensuring all the requirements of NAVSEA Standard Item 009-32 layout requirements are met. NAVSEA has a layout for nonskid application to rast track plates designated as Note-61.

FIGURE 6.15 Photo of applied nonskid to rast track plates off ship.

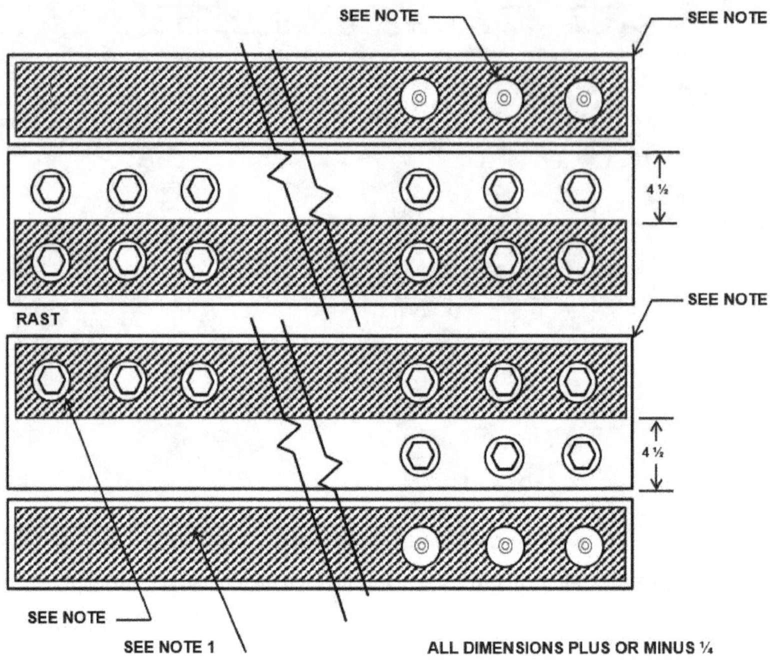

NOTES:
1. APPLY NONSKID ONLY TO CROSS-HATCHED AREAS.
2. NO NONSKID ON 2-3/4 to 3-1/4 INCH DIAMETER CIRCLES CENTERED ON HOLES.
3. ¾ INCH MASKING AROUND ALL PLATE EDGES.

FIGURE 6.16 NAVSEA Standard Item 009-32, Note-61, Nonskid layout requirements. Image and notes reprinted from the public domain document.

6.6.4 Applying Nonskid to Vertical Launch Systems (VLSs)

Applying nonskid to VLSs requires the nonskid color to be "haze gray"[1] vs. the traditional deck-gray color scheme. The entire VLS coating color is haze gray including the slip deck areas. This is the color that the Navy has delegated for VLS systems.

Applying nonskid to VLSs (Figures 6.17 and 6.18) has the same general requirements of all nonskid applications with the exception of the following:

- The VLS is composed of skirt and cell hatches.
- The nonskid is applied 2 in from edge of skirt and cell hatches.
- There is one specific change in the masking-off for nonskid application.
- The corners of the cell hatches are masked off at 45° angles following the shape of the cell hatch design, giving the cell hatches an octagon shape.

1. "Haze gray" is the actual name of the paint color from the manufacturer and the Navy.

FIGURE 6.17 Photo of VLS cell hatches masked off with corners at 45° angles for nonskid application.

FIGURE 6.18 Photo of haze gray nonskid being applied to VLS skirt.

6.6.5 Sprayable Type-4 Nonskid

Other than rollable nonskid, there is type-4 spray on nonskid. This application uses special equipment. Spray-on nonskid does not provide either an aggressive peak-to-valley surface or appearance, such as rollable nonskid does using a hard core, smooth, phenolic roller. Spray-on nonskid uses special pots and spray guns. It is imperative these pots are under constant agitation ensuring the aggregates are flowing through the line through the special gun used for application.

Sprayable nonskid has different requirements than traditional rollable nonskids:

- Sprayable nonskid spread rate range is 40 sqft per gallon minimum allowed requirement up to 60 sqft per gallon maximum allowed requirement.
- When spraying the nonskid, there should be an overlap of a minimum of 75% for optimum coverage.
- During application, if cratering is seen in the spray applied nonskid, then it is being applied too heavily.
- When applying spray on nonskid, it should be at a 45° angle facing away from the applicator. This is due to overspray, or loss of the nonskid from spraying. Because of this, it is best management practices to point the gun forward at a 45° angle in the direction of the sprayable nonskid application.

FIGURE 6.19 Photo of type-4 spray applied nonskid in process. Photo courtesy of Felix Perez.

There are two spray guns most commonly used for spray-on nonskid applications:

- Binks-7E2 Gun[2]
- Pentech-MG3001[3]

The spray tips used are a 3/8 in or 3-1/8 in size orifice for large particle or aggregated coatings systems. The spray gun is a conventional style gun that has a coating line and air line connected to the gun, where the volume of air during application can be controlled.

2. The Binks brand is owned by Carlisle Fluid Technologies, Scottsdale, AZ.
3. Pentech is a trademark of Pentech, Inc., Sarasota, FL

When ordering nonskid for spray application, the same formula can be used as the rollable nonskid. Instead of using the average spread rate for rollable nonskid, use a spread rate of 35 gal per sqft. An average of 45–50 spread rate will be obtained with sprayable nonskid.

There are two styles of nonskid spray pots: one is an open agitated pot (Figure 6.20) and one a closed, pressurized, agitated pot (Figure 6.21). Note that this pressurized pot is different from the pressurized sand blast pot described in Chapter 1.

FIGURE 6.20 Photo of Pentech open nonskid spray pot. Photo courtesy of Felix Perez.

FIGURE 6.21 Photo of a 5 gal Binks pressurized, closed, nonskid spray pot.

6.6.6 Peel-and-Stick Nonskid

Peel-and-stick nonskid is applied by removing the protective liner that covers the pressure-sensitive adhesive side of the peel-and-stick sheet. Once the liner is removed, the adhesive side is laid on the deck, firmly laying one side of the peel-and-stick sheet ensuring the adhesive is tightly adhered to the deck (Figure 6.22). Then, unrolling it flat and evenly against the deck in one direction ensures there are no creases or lumps caused by air pockets.

FIGURE 6.22 Photo of peel-and-stick application.

Unlike rollable nonskid and sprayable nonskid, peel-and-stick has no required spread rate, but knowing the square footage is still needed to ensure enough peel-and-stick is ordered to cover the deck.

Peel-and-stick comes in many different shapes and sizes. The most common sizes used on Navy ships are 3×4 and 4×5 sheets. When ordering peel-and-stick, the manufacturer sells it in rolls and identifies how much square footage each roll will cover. This may vary by manufacturer.

The most common peel-and-stick purchased for Navy ships is one roll of peel-and-stick that covers 120 sqft. Knowing the square footage allows the contractor to order accordingly, but extra rolls should always be ordered in case the peel-and-stick is applied incorrectly and/or if a crease or air bubble appears in it. If applied incorrectly, it would then have to be pulled and replaced with a new sheet.

NAVSEA Standard Item 009-26, Attachment-G, covers the requirements for peel-and-stick application:

- The corners shall be rounded, vs. straight-edge corners that can lift with foot traffic. The rounded edges tend to not lift as easily as do straight-edge corners.
- Most manufacturers that sell to U.S. Department of Defense (DOD) contractors sell the peel-and-stick with rounded corners, meeting NAVSEA requirements.

- Once the peel-and-stick is applied on the deck, the edge sealer (which literally seals the edges) must be applied. This helps to further protect it from water or other moisture from creeping under the peel-and-stick; it also prevents it from peeling up with constant foot traffic.
- The most common sealer used is Safety-Walk[4] edging compound from 3M. Each tube is 5 fluid ounces (U.S.) and can cover 100 linear feet.

It is important to order enough edge sealer for the peel and stick application.

6.6.7 Applying Thermal Spray Nonskid

NAVSEA has included thermal spray nonskid for LHD and LHA class ships in landing spots 2, 4, 5, 6, 7, and 9. NAVSEA has implemented Standard Item 009-124 for this process. This is a highly unique nonskid coating that has been introduced to LHD and LHA flight decks because of traditional rollable nonskid cracking and failing due to high heat signatures buckling the deck, and in turn, causing the traditional nonskid to crack and fail. Chunks of failed nonskid then become foreign object debris (FOD), which can get sucked into a jet engine. FOD flying out of an aircraft would cause damage to the aircraft or worse—the death of a pilot or personnel working the flight deck.

Thermal spray nonskid (TSN) application consists of a remote-controlled, manned, robotic machine, which applies the TSN (Figures 6.23 and 6.24). TSN consists of two materials—Al aluminum alloy and TI titanium alloy. The application process uses these alloys in wire form through the twin wire arc spray application process. The percentage of each alloy in the wire is what creates the chemical makeup of the TSN, i.e., 97% Al aluminum alloy wire and 3% TI titanium alloy wire.

FIGURE 6.23 Photo of the wire spools fed through the Thermion[5] box for TSN application.

4. Safety-Walk is a trademark of 3M, St. Paul, MN
5. Thermion is a registered trademark of Thermion Inc., Silverdale, WA.

FIGURE 6.24 Photo of robotic machine thermal spray application.

The thermal sprayed nonskid is accomplished in two passes—one pass where the thermal spray coating is applied flat ensuring coverage, and the second pass, which is accomplished differently. The applicators use a different technique of having the machine sputter and pulsate the thermal spray in droplets. This is what gives the thermal spray its peak-to-valley coefficient of a friction antislip surface. The peaks are not as pronounced as the rollable nonskid application (Figure 6.25).

FIGURE 6.25 Photo of applied thermal sprayed nonskid.

This is not the final step in the TSN application process. Thermal spray is a porous coating that needs to be sealed. Also, the TSN needs to match the traditional nonskid dark gray color scheme (Figure 6.26). The coating manufacturer has made a sealer and a dark gray color topping to match the color scheme of the surrounding traditional nonskid.

FIGURE 6.26 **Photo of dark gray color topping applied to the thermal spray nonskid. Photo courtesy of Jose M. Sanchez.**

The clear seal coat is a single-coat, water-thin viscosity coating that seals and penetrates through the TSN to the substrate, sealing the porous TSN.

The dark gray color topping is a single component, low-gloss polysiloxane that is more of a traditional viscosity but is still designed to be applied as a fairly thin coat. This thin coat will not affect or jeopardize the coefficient of friction of the TSN. The dark gray color topping imparts additional corrosion protection and of course, color. Along with dark gray, there are yellow, white (Figure 6.27), and red color toppings for the visual landing aids (VLA). These color toppings have to hold up to just as much heat signatures as the thermal spray nonskid.

FIGURE 6.27 Photo of visual landing aid (VLA) white color topping application to TSN on landing spots 7 and 9. Photo courtesy of Hugo Machuca.

These single-coat Polysiloxane color toppings can hold up to approximately 650°F, which is higher than most standard coatings, so they fair well for the most part. Eventually however, they are designed to be replaced during any maintenance planning. The clear sealer typically only is applied during the initial installation and then color coats are repaired and/or replaced as needed.

7 Nonskid Failures and Repair Processes

7.1 Nonskid Failures

Nonskid is a coating, and like all coatings, there will be failures. It is knowing the types of failures—and knowing how to either mitigate failures or repair them properly—that is critical. Nonskid is not like most coatings that are a flat, smooth, even surface. When dealing with nonskid failures that are general to a specific area of less than 10 sqft, they can undergo spot repair.

There are many different reasons why nonskid failures occur. Improper mixing can lead to a major failure—this was covered in Chapter 4 for polyamide anticorrosive primers and in Chapter 6 for nonskid application. This chapter will focus on the failures that can occur even when the nonskid systems are mixed and applied correctly.

7.1.1 Nonskid Cracking

Figure 7.1 shows a crack (a.k.a. a failure) that might look small in nature (the pen is there for size comparison), but after further inspection it can be seen how big this crack will grow. After probing the crack using a flat-edge scraper, the failure grows (Figure 7.2). Some failures start off small. It is up to the inspector to scrape the edges around the nonskid until it stops delaminating. After scraping as far as it will go, enough pressure must be put on the scraper to the borders of the failure to make sure the boarders are tightly adhered and will not give anymore.

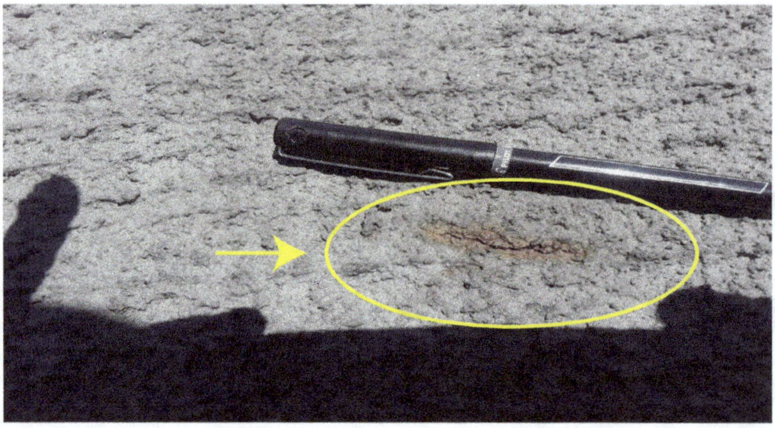

FIGURE 7.1 Photo of nonskid cracking on the USS *Essex* LHD-2 flight deck. Because it had only been in service for 30 days, this was unusual. The deck was inspected for further cracking, and none were found.

FIGURE 7.2 Photo of how a nonskid failure grows after flat-edge scraper probing.

7.1.2 Nonskid Repairs

Once probing is completed with the flat-edged scraper, there most likely will have an awkward shape left to repair. With other coatings, a contractor would be able to surface prepare the area as-is by feathering the edges and applying the new coating. However, because nonskid is a thicker, aggregated coating it will leave a jagged edge around its perimeter, so the edges cannot be feathered down. For nonskid repairs, the guideline is to extend the repair 1 ft outside the tightly adhered perimeter. This will ensure the failed area is mitigated; if it would continue to delaminate while power tooling, then a bigger problem would be found. This, however, was not the case. Here, the area was scraped until tightly adhered, and did not continue to delaminate on the USS *Essex* flight deck.

7.1.2.1 Repairs on Critical Coated Areas

For nonskid repairs on critical coated areas, a "conditioned found" report must be submitted to the U.S. Government requesting the contractor to use the appropriate power tools according to the SSPC-SP 11 standard. If the repair is small, the Government will usually approve the request with a caveat that the contractors ensure they have implemented the correct profile requirements.

As seen in Figure 7.3, the contractor went 1 ft out from the failure and squared up the repair using a handheld deck crawler using new deck-crawler teeth. This power tool provided the required profile. Squaring off the area makes it easy to apply the nonskid and makes it look uniform. Trying to apply nonskid to oval or other oddly shaped repair areas would not look aesthetically pleasing to the eye.

FIGURE 7.3 Photo of a contractor squaring off a 1 sqft area.

7.1.2.2 Repairs with Type-4 Sprayable Nonskid

Figure 7.4 shows a crack along the weld seam. This is a type-4 sprayable nonskid failure. In this type of nonskid, there is no cross-rolling of the welds, as seen in rollable nonskid applications. The cracking along the weld seam can be caused by many factors, e.g., applying sprayable nonskid too heavily along the weld seam and/or a breach in the barrier coat (primer), which can allow in moisture and would cause the crack.

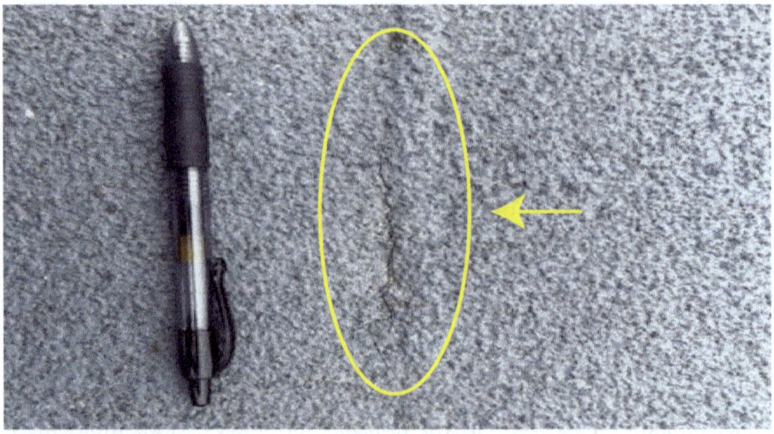

FIGURE 7.4 Photo of a crack along a weld seam. Photo courtesy of Hugo Machuca.

7.1.2.3 Repairing Type-4 Nonskid with Isolated Failures

Sprayable nonskid has a different texture and appearance vs. rollable nonskid. As such, a nonskid spray pot for a spot repair would not be used—it would be excessive, and having overspray on existing nonskid is bad. It could become foreign object debris (FOD).

Spot repair of sprayable nonskid will be similar to repairing rollable nonskid as far as surface preparation and primer application, with the exception of the nonskid application process itself. In this case, the contractor would mix a kit of type-4 nonskid, but instead of using a hard, smooth, phenolic roller, a roller nap is used. When using the roller nap, aggregates on the nap must be ensured. Also, too much pressure on the roller nap cannot be applied so that an even dispersal of the aggregated coating matching the surrounding nonskid can be ensured.

7.1.2.3.1 Microchecking in Nonskid

Microchecking are tiny fractures at the surface of the nonskid (Figures 7.5 and 7.6) that do not go to the barrier coat or substrate; in nonskids, it is most common with polysiloxane nonskids. Epoxy nonskids also have been known to have microchecking. If the microchecking is only at the surface of the nonskid, and/or if it was to a general area that was not big or was not throughout the nonskid, it would not be considered a failure.

The nonskid microchecking area would be monitored to see if it continued or if it went down to the barrier coat or substrate. If microchecking was throughout the entire area of the nonskid application, this would then be considered a failure because it is throughout and is not worth jeopardizing aircraft or ship's personnel.

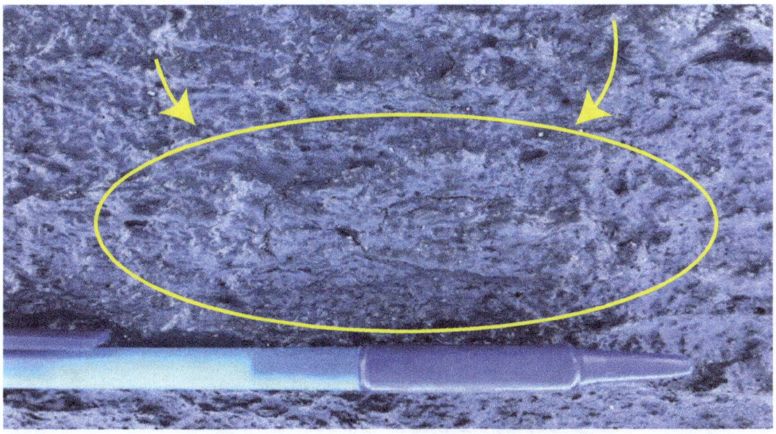

FIGURE 7.5 Photo of newly applied nonskid microchecking. Photo courtesy of Tomas Ysidro.

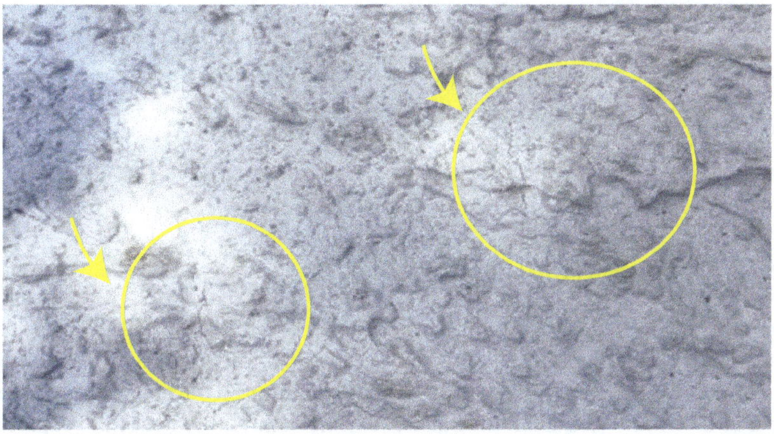

FIGURE 7.6 Photo of microchecking on nonskid that has been down for 30 days. Photo courtesy of Tomas Ysidro.

7.1.2.3.2 Microcracking in Nonskid

Microchecking has led to microcracking (Figure 7.7), which then went down to the barrier coat and the substrate—this would automatically be a failure. The way to verify if microchecking has led to microcracking is by taking a dull putty knife and applying pressure to the microchecking areas. If they do not crack or delaminate, this would be just microchecking. If the dull putty knife does cause the microchecking to crack to the barrier coat and substrate, then it would be microcracking, which is now a nonskid failure.

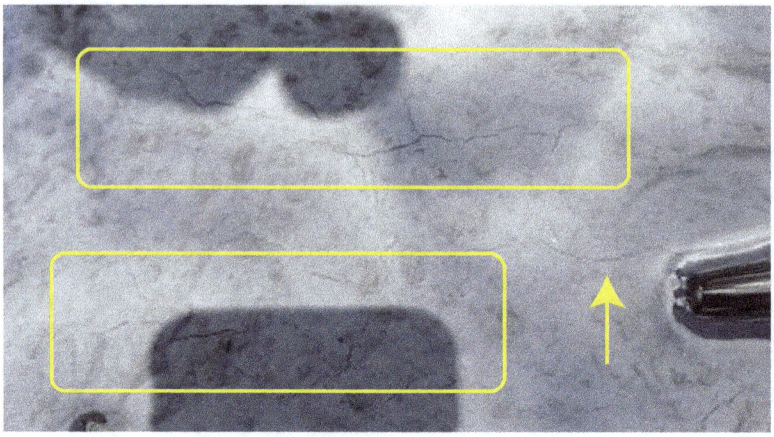

FIGURE 7.7 Photo of microcracking, which is more pronounced and wider than microchecking. Photo courtesy of Tomas Ysidro.

If nonskid microcracking is found in a general area of less than 10 sqft and not throughout, the entire nonskid system is applied—the same method can be used of using a dull putty knife until tightly adhered. The localized area can be squared off, as seen in the previous repair procedure, for small areas of failed nonskid.

7.1.2.4 Thermal Spray Nonskid (TSN) Repair Process

There is a different repair process for small areas of failures. Just like rollable and sprayable nonskid, TSN failures start with cracking (Figure 7.8). Unlike rollable and sprayable nonskid, however, the repair process is more complex. For TSN application, the minimum surface preparation requirement to follow is the SSPC-SP 5, "White Metal Blast Cleaning" standard. For small repairs, the thermal spray robotic machine is not used because it is not cost effective. The contractors prefer to use a handheld thermal spray wand (Figures 7.9 and 7.10).

FIGURE 7.8 Photo of a crack in thermal spray nonskid. Photo courtesy of Gordon Kuljian.

FIGURE 7.9 Photo of thermal spray nonskid application using a handheld thermal spray wand. Photo courtesy of Gordon Kuljian.

FIGURE 7.10 Photo of a thermal spray nonskid spot repair. Photo courtesy of Gordon Kuljian.

There are nonskid failures in specific locations that cannot be squared off and repaired because of the critical nature of its service. A CVN flight deck wire box area is an area that traps aircraft with four arresting gear cables across the deck port to starboard; the aircraft uses a tailhook that can catch one of the four cables—this is known as an arrest or trap. When the tailhook hits the nonskid it should leave a straight line or gash removing the nonskid to bare metal (Figure 7.11). If a tailhook hits the nonskid and it leaves an oval or round shape to bare metal this would be considered a failure known as "pancaking" (Figure 7.12).

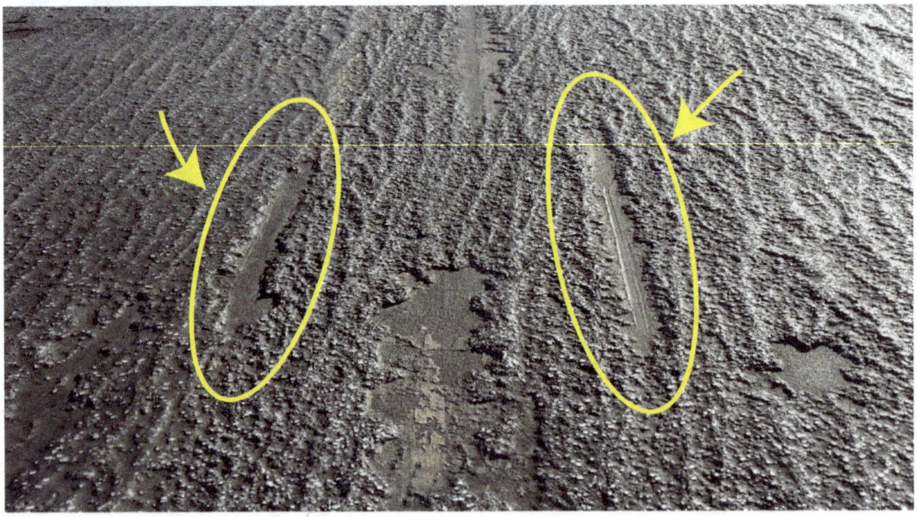

FIGURE 7.11 Photo of acceptable tailhook damage to Comp-L limited nonskid.

FIGURE 7.12 Photo of a tailhook impact causing pancaking of the Comp-L nonskid—this is not acceptable. Pancaking is the nonskid delaminating in an oval or circular shape off of the substrate from the impact of the tailhook. The nonskid should be tightly adhered only leaving a straight line or streak to bare substrate, as seen in the previous photo (Figure 7.11).

Author's Final Notes

Nonskid is an important asset for the U.S. Navy—it is a safety antislip coating that helps protect Navy personnel, aircraft, and vehicle movements when at sea. For all contractors and coatings inspectors who plan on taking a Navy ship nonskid contract, it is your duty to ensure you meet all requirements of NAVSEA Standard Item 009-32. Most importantly however, nonskid should be applied with the integrity of the contractor and inspectors ensuring that our Navy's ships go out to sea feeling confident that the nonskid they walk on, fly on, and make any kind of movement on, gives them the security in knowing they can operate safely when at sea. Something my mentor, Bernard Maertz, always taught me was "our main goal is to protect our Navy".

References

The publication information for the standards cited in this book can be found below under the names of their respective publishers. In some cases, the standard number does not correspond to the publisher's name, such as with Naval Sea Systems Command publishing "MIL" standards, or AMPP publishing "NACE" and "SSPC" standards.

Note that because of the NACE–SSPC merger, all NACE, SSPC, and joint NACE/SSPC and NACE/SSPC standards cited in the book are listed under AMPP. Likewise, joint ANSI/NACE standards are also listed under AMPP.

American Society for Testing and Materials (ASTM)

ASTM D1640, "Standard Test Methods for Drying, Curing, or Film Formation of Organic Coatings," 2022, publication version used: D1640_D1640M-14R22.

ASTM D4285-83, "Standard Test Method for Indicating Oil or Water in Compressed Air," 2018.

ASTM D4940-15, "Standard Test Method for Conductimetric Analysis of Water Soluble Ionic Contamination of Blast Cleaning Abrasives," 2020, publication version used: D4940-15R20.

ASTM D7393-16, "Standard Practice for Indicating Oil in Abrasives," 2020, publication version used: D7393-16R20.

ASTM F21-20 "Standard Test Method for Hydrophobic Surface Films by the Atomizer Test," 2020, publication version used: F0021-20.

ASTM E11-22, "Standard Specification for Woven Wire Test Sieve Cloth and Test Sieves," 2022.

ASTM F22-21, "Standard Test Method for Hydrophobic Surface Films by the Water-Break Test," 2021, publication version used: F0022-21.

ASTM F718, "Standard Specification for Shipbuilders and Marine Paints and Coatings Product/ Procedure Data Sheet," 2022, publication version used: F0718-22.

Association for Materials Protection and Performance (AMPP, formally SSPC and NACE)

ANSI/NACE SP0508-2017, "Methods of Validating Equivalence to ISO 8502-9 on Measurement of the Levels of Soluble Salts," 2017.

SSPC-AB-1-2019, "Mineral and Slag Abrasives," 2019.

SSPC-AB-2-2016, "Cleanliness of Recycled Ferrous Metallic Abrasives," 2016.

SSPC-AB-3-2017, "Ferrous Metallic Abrasive," 2017.

SSPC Guide 15-2020, "Field Methods for Retrieval and Analysis of Soluble Salts on Steel and Other Nonporous Substrates," 2020.

SSPC Guide 24-2018, "Soluble Salt Testing Frequency and Locations on New Steel Surfaces," 2018.

SSPC-SP 1, "SSPC-SP 1-2016, Solvent Cleaning" 2016.

SSPC-SP 5 (Joint NACE standard #SSPC-SP 5/NACE No. 1-2006), "White Metal Blast Cleaning," 2006.

SSPC-SP 10/NACE No. 2-2006, "Near-White Metal Blast Cleaning," 2006.

SSPC-SP 11, "Power Tool Cleaning to Bare Metal," 2020.

SSPC-SP 17, "Thorough Abrasive Blast Cleaning of Non-Ferrous Metals," 2019.

SSPC-SP WJ-2-2021 (Joint NACE standard #NACE WJ-2/SSPC-SP WJ-2-2021), "Waterjet Cleaning of Metals—Very Thorough Cleaning (WJ-2)," 2021.

SSPC-SP WJ-3-2021 (Joint NACE standard #NACE WJ-3/SSPC-SP WJ-3-2021), "Waterjet Cleaning of Metals—Thorough Cleaning (WJ-3)," 2021.

SSPC-SP WJ-4-2021 (Joint NACE standard #NACE WJ-4/SSPC-SP WJ-4-2021), "Waterjet Cleaning of Metals—Light Cleaning (WJ-4)," 2021.

SSPC-VIS 1, "Guide and Reference Photographs for Steel Surfaces Prepared by Dry Abrasive Blast Cleaning," 2012.

SSPC-VIS 3, "Guide and Reference Photographs for Steel Surfaces Prepared by Power and Hand Tool Cleaning," 2007.

SSPC-VIS 4 (Joint NACE standard #SSPC-VIS 4/NACE-VIS 7), "Guide and Reference Photographs for Steel Surfaces Prepared by Waterjetting," 2005.

SSPC-WJ-1 (Joint NACE standard #NACE WJ-1/SSPC-SP WJ-1), "Waterjet Cleaning of Metals—Clean to Bare Substrate (WJ-1)," 2021.

International Organization for Standardization (ISO)

ISO 8502-3, "Preparation of Steel Substrates Before Application of Paints and Related Products. Tests for the Assessment of Surface Cleanliness Part 3: Assessment of Dust on Steel Surfaces Prepared for Painting (Pressure-Sensitive Tape Method)," 2017.

ISO 8502-6, "Preparation of Steel Substrates Before Application of Paints and Related Products. Tests for the assessment of Surface Cleanliness Part 6 Extraction of water soluble contaminants for analysis (Bresle Method)," 2020, publication version used: Edition 3.

ISO 8502-9, "Preparation of Steel Substrates Before Application of Paints and Related Products. Tests for the assessment of Surface Cleanliness Part 9: Field method for the conductometric determination of water-soluble salts," 2020, publication version used: Edition 2.

ISO 8502-10, "Preparation of Steel Substrates Before Application of Paints and Related Products. Tests for the assessment of surface cleanliness Part 10: Field method for the titrimetric determination of water-soluble chloride," 1999, publication version used: Edition 1.

Naval Sea Systems Command

MIL-A-22262, "Abrasive Blasting Media Ship Hull Blast Cleaning," 2021, publication version used: Revision B Notice 1.

MIL-A-A-1722, "Grain, Abrasive (Soft Blasting)," 2017, publication version used: Notice 3.

MIL-A-A-59316, "Commercial Item Description Abrasive Materials; for Blasting," 1999.

MIL-PRF-24667, "Coating System, Nonskid, for Roll or Spray Application," 2021, publication version used: Revision D.

MIL-PRF-81733 Type-2, "Sealing and Coating Compound, Corrosion Inhibitive," 2019, publication version used: Revision D Notice 1, Validation.

NAVSEA Standard Item 009-32 (FY-23), "Cleaning and Painting Requirements; accomplish," 2021, publication version used: FY-23.

NAVSEA Standard Item 009-32 Note-61, "Nonskid Layout Requirements: Cleaning and Painting Requirements; accomplish," Note 61, 2022, publication version used: FY-24.

NAVSEA Standard Item 009-124, "Thermal Spray Nonskid Application; accomplish," 2022, publication version used: FY-24.

NSTM Chapter 634, "Naval Ship's Technical Manual Deck Coverings, General," 2009, publication version used: Volume 1, Revision 4.

NAVSEA Standard Item 009-32 Nonskid Notes

NAVSEA 009-32 Standard Item is a category 2 document. NAVSEA has category 1 standard items that are mandatory to follow and are included in all NAVSEA projects. Category 2 standard items such as 009-32 must be identified in the specification if it is to be invoked on the project.

The standard has notes that are contingencies, and some may give options, which can be invoked by requesting Government permission. Some notes are mandatory for nonskid systems. These notes are specific to the U.S. Government's Fiscal Year (FY) 2023. Please be sure to reference the current FY standard for a given project to ensure all notes are correct. NAVSEA updates NAVSEA Standard Item 009-32 annually.

Except for the footnote at Note-61, these notes were taken directly from 009-32 and have not been altered.

Note-7. Within a zone or work area, a proprietary nonskid system listed on the QPL for MIL-PRF-24667 must be used.

Note-16. Apply polysulfide compound conforming to MIL-PRF-81733, Type II, Class 2, after primer application, prior to topcoat application.

Note-19. MIL-PRF-24667 nonskid systems must be applied as a complete system from the same manufacturer. The use of any manufacturer's MIL-PRF-24667 color topping is authorized over any MIL-PRF-24667 nonskid including type XI nonskid.

Note-22. Peripheral deck edging and areas not receiving nonskid may use either the manufacturer's MIL-PRF-24667 color topping or MIL-PRF-24635 Type V. Aircraft tie-downs may be coated with MIL-DTL-24441 F-155 in lieu of nonskid color topping.

Note-50. "Cosmetic" color topping is not to be applied on top of nonskid on vertical replenishment or aviation decks.

Note-56. Do not apply nonskid to 7-inch-wide strip of deck surface in way of the hangar door seal interface on DDG-51 Class Flight II-A ships.

Note-59. Composition L nonskid may be used in Composition G areas with TYCOM approval.

Note-60. MIL-PRF-24667 Composition L material must be applied in areas designated by the VLA General Services Bulletin Number 8.

Note-61. For flight decks equipped with RAST tracks, see Figure One for guidance on RAST track plates.[1]

Note-62. Once the old nonskid system is removed, or if it is a new deck being prepared for nonskid system installation, the surface of the GRP fiberglass must be lightly hand-sanded with 80-grit paper to roughen the surface.

Note-63. Nonskid must be applied to within 2-3 inches of deck fittings and protrusions. Nonskid must be applied to within 4-6 inches from coamings and deck edges. Areas underneath protrusions, such as stowage racks or fire stations that are normally found on hangar decks, or coaming areas of decks not subject to vehicle or aircraft movement, need not comply with the 4-6 inch rule for applying nonskid near deck edges.

Note-72. MIL-PRF-24667 Type I spray-applied products are not authorized for application to critical coated areas on CVNs.

Note-75. The entire deck surface (i.e., forecastle, fantail, flight, etc.) containing a Vertical Replenishment Deck area must be considered a critical coated area as addressed in 3.7.

Note-76. For aircraft carriers, substitute 1 stripe coat of primer for the 2nd primer full coat.

Note-80. Maximum water pressure must be such that 30 seconds of stationary operation of the vacuum self-contained equipment produces no visible degradation of the nonskid surface. Vacuum self-contained equipment must be capable of cleaning within 4 inches of bulkheads, coamings, and other deck appurtenances.

Note-82. Each cell hatch lip must be prepared to SSPC-SP 2. (**Vertical Launch System**)

Note-83. Nonskid on vehicle ramps must be rolled perpendicular to the main axis of the ramp. Welds must not be cross-rolled on vehicle ramps.

Note-86. For aircraft carriers only, MIL-PRF-24635, Type III, Class 2, Grade B or C may be substituted where MIL-PRF-24635, Type V and VI, Class 2, Grade B or C is specified.

Note-89. When hand tool or power tool aluminum cleaning to SSPC-SP 2 Hand Tool Cleaning, SSPC-SP 3 Power Tool Cleaning, SSPC-SP 11 Bare Metal Power Tool Cleaning, and SSPC-SP 15 Commercial Power Tool Cleaning only stainless-steel needles, stainless steel bristle impact belts, stainless steel wire brushes, stainless steel scouring pads, or aluminum oxide abrasive sanding pads must be used. None of these materials must have been previously used on other metal, or for the removal of copper pigmented paint, prior to being used on aluminum. Rotary sanders or grinders fitted with abrasive discs or flap wheels, and rotary impact tools fitted with cutter or hammer bundle assemblies are not authorized on aluminum substrates. Rotary impact tools may only be fitted with tungsten carbide tipped, rubber impregnated woven nylon cloth flaps.

---

1. A drawing of rast track plates are included in Chapter 6 of this book.

Note-90. When water jetting an aluminum surface to SSPC-SP WJ-2/NACE WJ-2, the surface cleanliness requirements must be the same as that stated in the standard, except that the words “corrosion products” must be substituted for the word “rust”; and the flash rusting requirement does not apply.

Note-91. For nonskid removal using water jetting on LCS 2 Class ships, the pressure must not exceed 30,000 PSI.

Note-92. After nonskid installation, remove masking from bolted-crossbar aircraft securing fittings. Unless directed by the SUPERVISOR, do not conduct surface preparation or apply coatings to bolted-crossbar aircraft securing fittings.

Note-44A. For Submarines-Do not apply nonskid to Fairwater Planes or Retractable Bow Planes.

Naval Ship Deck Square Footages for Each Class of Ship

B.1 Introduction

This appendix lists the deck square footages taken by the author during his 30-year career in quality assurance and quality control (QA/QC) in nonskid coatings. These measurements are required to calculate the nonskid spread rate per NAVSEA requirements. Knowing these square footages is vital to bidding on contracts because nonskid is bid by dollars per square foot.

Each table is structured the same way. They list the square footages for each class of U.S. Navy ship in terms of measurements for various decks and rooms found aboard these ships. "BLAST" refers to the square footage that is accessible to blast equipment to remove the nonskid; "POWER TOOL" refers to the square footage of the deck that are inaccessible to blast equipment so only power tools can be used to remove the nonskid and coatings. Note that certain letter combinations in the tables are U.S. Navy designations for various things and do not have actual definitions.

TABLE B.1 Square footages for Frigate, Guided missile class type ships (FFGS).

SHIPS' MEASUREMENTS FFGS	BLAST	POWER TOOL	TOTAL SQUARE FEET
FLIGHT DECK	3,671	641	4,312
FLIGHT DECK RAST	3,671	641	4,312
RAST TRACK PLATES HANGER BAY PORT AND STARBOARD (H/B P/S)		280	280
POOP DECK		370	370
HANGAR BAY P/S	760	210	970
HANGAR BAY P/S	814	203	1,017
HANGAR BAY P/S RAST	570	400	970
HANGAR BAY P/S SIDES RAST	610	407	1,017
RAST TRACK PLATES FLIGHT DECK P/S		455	455
MAIN DECK P/S	888	380	1,268
MAIN DECK P/S	888	380	1,268
FORECASTLE	2,538	1,188	3,726
BOAT DECK P/S		450	450
02 LEVEL	3,712	3,709	7,421
03 LEVEL	429	531	960
PASSAGEWAY (01-212-0-M)		233	233
TORPEDO MAGAZINE		390	390
CENTER LINE PASSAGEWAY	582	388	970
TACTICAL STOWED ARRAY SONAR (TACTAS) BATHY ROOM		329	329
76 MM (1-220-1-T)		200	200
ELEVATOR PLATFORM (4-203-0-T)		100	100
HELICOPTER DECK WORKSHOP		213	213
(3-188-0-L)		300	300
TRUNK 2-301-1-T		36	36

TABLE B.2 Square footages for Cruiser Guided (CG) missile type ships.

CG'S TICONDEROGA CLASS 52-73	BLAST	POWER TOOL	TOTAL SQUARE FEET
FLIGHT DECK	2,165	1,308	3,473
6" TIE DOWNS	145		
FANTAIL	1,969	1,242	3,211
FORECASTLE	3,987	1,709	5,696
PORT AND STARBOARD (P/S) BREEZE EACH SIDE		216	216
02 LEVEL BOAT DECK P/S		385	385
MISSILE DECK 01 PASSAGEWAY P/S	3,336	1,797	5,133
MISSILE DECK ONLY	1,120	536	1,656
01 P/S FORWARD OF MISSILE DECK	2,334	1,000	3,334

PASSAGEWAY P/S (01-138-2/3-L)	260	172	432
02 LEVEL BOAT DECK P/S		464	464
LOADING PLATFORM P/S		66	66
02 LEVEL CATWALK P/S	113	228	341
HELICOPTER DECK HANGAR	1,187	640	1,827
6" TIE DOWNS	100		
HELICOPTER DECK REPAIR SHOP		180	180
JP-5 FUELING STATION		100	100
RAST LSO CONTROL STATION		400	400
03 LEVEL	1,099	732	1,831
04 LEVEL FORWARD	1,944	487	2,431
04 LEVEL AFT	409	409	818
05 LEVEL		300	300
05 LEVEL AFT		521	521
05 LEVEL AFT OE-82 PLATFORM		84	84
06 LEVEL FORWARD		791	791
06 LEVEL AFT		650	650
MAIN MAST PLATFORM		356	356
FORE MAST YARDARM AND PLATFORM		260	260
PASSAGEWAY (1-426-1-L)		112	112
PASSAGEWAY (1-382-01-L)		324	324
TORPEDO ROOM # 1 (1-390-1-M)		450	450
TORPEDO ROOM # 2 (1-390-4-M)		450	450
SECURITY STA. # 2 (1-464-2-Q)		25	25
SONOBUOY STORAGE ROOM 01-398-1-A		200	200
PASSAGEWAY (2-382-3-L)		78	78
5/54 AMMUNITION ROOM #1 (2-58-0-Q)		450	450
5/54 AMMUNITION ROOM #2 (3-464-0-Q)		308	308
NIXIE ROOM		200	200
TACTICAL TOWED ARRAY SONAR (TACTAS) 2-506-0-Q		229	229
TRUNK (2-506-2-T)		32	32
TRUNK (3-127-2-T)		60	60
PASSAGEWAY (1-58-01-L)		391	391
TRUNK (3-58-0-T)		20	20
PASSAGEWAY (1-426-1-L)		145	145
BOW + STARBOARD HAWSE PIPE, CHAIN PIPE BOLTER, BITTS, CHOCK CLOSED, CHOCK ROLLER		2,540	2,540
TOWED ARRAY ROOM (2-506-0-Q)		183	183
FANTAIL	2,370	790	3,160
MISSLE DECK	1,304	558	1,862
01 LEVEL PASSAGEWAY P/S	2,280	976	3,256

01 LEVEL BREAKS	260	112	372
FOCSLE	4,028	1,007	5,035
LIGHT TRAP FRAME 308-312		32	32
LIGHT TRAP FRAME 410-418		21	21
LIGHT TRAP FRAME 346-356		46	46
LIGHT LOCKER 03-174-0-L		44	44
PASSAGEWAY ATHWARTSHIP 1-382-01-L		342	342
PASSAGEWAY 03-174-0-L		355	355
PASSAGEWAY 03-289-2-L		76	76
PASSAGEWAY 04-292-2-L		108	108
RAST MACH ROOM 2-442-0-E		374	374

TABLE B.3 Square footages for Destroyer (DD) Spruance Class ships.

SHIPS' MEASUREMENTS DESTROYER (DD) SPRUANCE CLASS	BLAST	POWER TOOL	TOTAL SQUARE FEET
BOAT DECKS PORT AND STARBOARD (P/S)		468	468
HELICOPTER DECK (02 LEVEL)	2,436	609	3,045
HANGAR DECK (02 LEVEL)	1,190	297	1,487
BOAT DECK P/S		468	468
03 LEVEL MID SHIP	841	915	1,756
03 P/S BRIDGE WINGS		470	470
04 FORWARD SIGNAL BRIDGE	1,166	1,166	2,332
04 AFT	1,365	909	2,274
05 FORWARD CLOSE-IN WEAPON SYSTEM (CIWS) GUN DECK		192	192
06 AFT CIWS GUN DECK		252	252
FORWARD MAST		698	698
MAIN MAST		753	753
HORSESHOE PASSAGEWAY 1-382-01-L		300	300
STARBOARD TORPEDO MAGAZINE 1-390-1-M		420	420
PORT TORPEDO MAGAZINE 1-390-2-M		420	420
AFT PALLET STAGING 3-464-01-Q		121	121
5/54 PROJECTILE 3-62-0-M		450	450

TABLE B.4 Square footages for the Missile Guided Destroyer (DDG) Milius class 51 to 78 ships.

DDG MILIUS CLASS 51 TO 78	BLAST	POWER TOOL	TOTAL SQUARE FEET
FLIGHT DECK	3,873	969	4,842
MILIUS FOSCLE	5,433	1,358	6,791
PORT BREEZE WAY		616	616
STARBOARD BREEZE WAY		616	616
PORT WEATHER	244	61	305
STARBOARD BOAT DECK	900	225	1,125
MISSLE DECK	2,657	665	3,322
PORT AND STARBOARD (P/S) ARRAY DECKS	851	213	1,064
P/S MID QUARTER DECK	1,410	352	1,762
02-242-4-WEA (WEATHER DECK)	775	194	969
02 LEVEL STARBOARD SIDE		438	438
02 LEVEL CENTER SECTION		220	220
03-163-2-WEA		114	114
03-178-0-WEA		442	442
03-178-1-WEA		442	442
3-42-0-M		404	404
STARBOARD 04 LEVEL BRIDGE WING		320	320
PORT 04 LEVEL BRIDGE WING		320	320
04-285-0-WEA DIR (WEATHER DIRECTOR DECK)		168	168
05-272-0-WEA DIR		225	225
UNREP STATION 3		231	231
UNREP STATION 4		315	315
STARBOARD COUNTERMEASURES CHAFF AND DECOY LAUNCHING SYSTEM (SRBOC) 02-205-1-WEA DECK		218	218
PORT SRBOC 02-205-2-WEA		218	218
P/S 05 LEVEL 05-130-1-WEA		826	826
P/S BRIDGE WINGS		600	600
CLOSE-IN WEAPON SYSTEM (CIWS) MOUNT 21 DECK		166	166
3 GTG DECK		391	391
GUN MOUNT ROOM (MK) 82 04-295-C-L		145	145
MK 82 MOD 05-282-CL		145	145
U-SHAPED DECK MOUNT 22		156	156
TACTICAL TOWED ARRAY SONAR (TACTAS) EQUIPMENT ROOM 2-506-0-Q		229	229

TABLE B.5 Square footages for the Missile Guided Destroyer's (DDG) Arleigh Burke Class 85 to 103 ships.

DDG'S ARLEIGH BURKE CLASS 85 TO 103	BLAST	POWER TOOL	TOTAL
FLIGHT DECK	3,459	865	4,324
05-285-0-WEA DIR (WEATHER DIRECTOR) DECK		225	225
HELICOPTER DECK FUELING STATION 1-399-0-J		106	106
HELICOPTER DECK HANGAR 1 RAST PLATE	853	365	1,218
HELICOPTER DECK HANGAR 2 INC RAST PLATE	853	365	1,218
RAST MACHINARY ROOM 2-442-0-E		374	374
PORT SIDE MAIN DECK	963	241	1,204
STARBOARD SIDE MAIN DECK	1,265	317	1,582
FOSCLE	5,193	1,298	6,491
MAIN DECK AMID SHIP	530	133	663
PASSAGEWAY TUNNEL 01-118-1-L	550	137	687
PASSAGEWAY TUNNEL 01-118-4-L	550	137	687
02 LEVEL FRAME 194-238 P/S	4,649	3,270	7,919
04-297-0-WEA DIR (WEATHER DIRECTOR) DECK		168	168
03-178-WEATHER ARRAY DECK	917	393	1,310
#3 REFUEL STATION		310	310
#4 + ½ REFUEL STATION DECK		320	320
02.5-372-WEA TOP NR3 GTG TRUNK		279	279
UPPER BOAT DECK		78	78
BOAT DAVIT PLATFORM		12	12
PORT BRIDGE WING 04-130-2 WEA		146	146
STARBOARD BRIDGE WING 04-130-1 WEA		146	146
FORWARD VERTICAL LAUNCH SYSTEM PERIMETER		268	268
SIGNAL BRIDGE		161	161
NIXIE ROOM		240	240
RAST MACH ROOM 2-442-0-E		374	374

TABLE B.6 Square footages for amphibious assault landing helicopter deck (LHD) ships.

AMPHIBIOUS ASSAULT SHIPS LHD	BLAST	POWER TOOL	TOTAL SQUARE FEET
FLIGHT DECK WITH ELEVATORS	92,500	3,000	95,500
EACH ELEVATOR	1,880	470	2,350
CAT WALKS P/S	2,240	2,240	4,480
TROOP WALKWAYS S/S 13-110 AND 121-130	2,326	2,325	4,651
TROOP WALKWAYS P/S 13-85 AND 97-130	2,313	2,313	4,626
TOP OF RAMP F/D	313	135	448
RAMP FLIGHT DECK TO HANGAR BAY	936	234	1,170
RAMP HANGAR BAY TO UPPER "V"	888	222	1,110
RAMP PIER TO UPPER "V"		573	573
FORWARD MAST PLATFORM 155 FT ABOVE BASELINE (ABL) AND 166		100	100

FORWARD MAST		500	500
AFT MAST		1,670	1,670
HANGAR BAY	19,889	4,419	24,308
ELEVATOR APPROACH EACH	525	225	750
04 LEVEL FORWARD MISSLE	1,096	470	1,566
02 LEVEL AFT MISSILE FRAME 130	720	308	1,028
FOCSLE		3,338	3,338
FANTAIL	875	938	1,813
FUEL SEA STATION #2	276	276	552
REP. SEA STATION 01 P/S		200	200
05 LEVEL PLATFORM FRAME 89-93		860	860
04 LEVEL LAUNCH PLATFORM 53-65	1,100	900	2,000
06 LEVEL FRAME 67-85	1,732	956	2,688
07 LEVEL AFT FRAME 87	819	410	1,229
07 LEVEL FORWARD FRAME 65-72	947	406	1,353
07 LEVEL DIRECTOR PLATFORM FORWARD		120	120
07 LEVEL DIRECTOR PLATFORM AFT		216	216
08 LEVEL FORWARD PLATFORM		154	154
08 LEVEL DIRECTOR PLATFORM AFT		300	300
09 LEVEL AFT RADAR PLATFORM		800	800
AFT MISSILE DECK	948	237	1,185
RHIB DK		428	428
02-113-01-L INSIDE WEATHER DECK 02-121-8-DK		79	79
02-113-01-L INSIDE WEATHER DECK 02-130-2-DK		35	35
SCUPPERS AND COMINGS	6,500	739	7,239
PLATFORMS TO FLT DK		450	450
ACCESS 01-88-1-L		116	116
01 LEVEL STOWAGE PLATFORM IN HB 93-97		379	379
SHIPS ARMORY 1-65-5-Q		299	299
ACCESS 1-88-1-L		91	91
ACCESS 02-13-7-L		16	16
NSSMS ROLLING AIRFRAME MISSILE (RAM) HAND EQUIPMENT STORAGE ROOM 1 04-65-5-A		147	147
SIDE WINDER CHARGING STATION BOTTLE STORAGE 03-57-1-Q		150	150
MOBILE MAINTENANCE FACILITY INTER ROOM 02-121-1-Q		800	800
PASSAGEWAY 1-67-1-L	300	981	1,281
WELL DECK VENT TRUNK 1-73-3-T		542	542
AVIA GROUND SUPPORT EQUIPMENT STOR-AGE 1-78-0-Q	600	400	1,000
MOORING STATION 2-54-4-Q		273	273
MOORING STATION 2-105-4-Q		273	273

TABLE B.7 Square footages of Amphibious Landing Helicopter Assault (LHA) ships.

AMPHIBIOUS ASSAULT SHIPS LHA'S	BLAST	POWER TOOL	TOTAL SQUARE FEET
FLIGHT DECK WITH ELEVATORS	81,574	4,293	85,867
DECK EDGE ELEVATOR	1,305	326	1,631
AFT ELEVATOR	1,632	408	2,040
HANGAR BAY	17,500	1,943	19,443
HANGAR BAY HAZARDOUS MATERIALS AREA	1,865	800	2,665
AIRCRAFT SHOP (1-122-1-Q)	996	426	1,422
HANGAR BAY DECK EDGE ELEVATOR APPROACH	519	222	741
CAT WALKS, PORT AND STARBOARD (P/S)		2,927	2,927
CAT WALKS P/S		1,721	1,721
TROOP WALK / SPONSONS P/S	2,724	1,861	4,585
TROOP WALK / SPONSONS P/S	2,578	1,719	4,297
TOP OF RAMP (03-58-1-Q)	712	305	1,017
RAMP FLIGHT DECK TO HANGAR BAY	2,176	544	2,720
RAMP HANGAR BAY TO UPPER "V"	900	225	1,125
RAMP PIER TO UPPER "V"		573	573
RAMP TO WELL (NO WELL)		2,838	2,838
LOWER "V" FORWARD	3,295	1,412	4,707
LOWER "V" AFT (STOREROOM)	3,738	1,602	5,340
UPPER "V"	13,385	5,736	19,121
WELL DECK RAMPS P/S (BARS)		2,520	2,520
HERO'S ALLEY	294	196	490
CLOSE-IN WEAPON SYSTEM (CIWS) P/S AFT		495	495
CIWS P/S AFT		421	421
AFT RAM	221	221	442
CIWS FORWARD ON ISLAND		400	400
02-N-P (SPONSON)	517	345	862
02-N-S (SPONSON)	517	345	862
02-125-6-M (MAGAZINE)		100	100
1-81-3-L PASSAGEWAY		140	140
07 LEVEL WEATHER	802	802	1604
SIGNAL BRIDGE (08-65-1-C)	3,974	1,703	5,677
RAIN GUTTERS/ TROUGHS		2,662	2,662
AVIATION READY ROOM (03-56-1-M)	295	197	492
TRIAGE (03-73-3-Q)	643	161	804
TRIAGE (03-81-1-Q)	901	225	1,126
FLIGHT DECK LOUNGE	107	71	178
BALLOON ROOM (03-86-1-Q)	176	117	293
PASSAGEWAY (1-73-2-L)		210	210
PASSAGEWAY (1-81--3-L)		180	180
PASSAGEWAY (03-83. 5-1-L)		108	108
PASSAGEWAY (03-78. 5-1-L)		33	33

PASSAGEWAY (03-79-3-L)		108	108
MOORING STATION (1-77-3-Q)		174	174
FORWARD CIWS MAGAZINE (04-56-1-M)		96	96
RADAR DECK (04-63-1)	978	652	1,630
AFT MAST		1,156	1,156
FORWARD MAST		1,104	1,104
BRIDGE CRANE WALK		1,020	1,020
ACCESS (1-92-2-L)		83	83
NITROGEN ROOM		215	215
ACCESS (1-120-1-L)		91	91
NIXIE ROOM (2-125-1-Q)		310	310
NIXIE ROOM (2-125-2-Q)		310	310
ANTENNA PLATFORM 06 LEVEL FRAME 89 AFT		912	912

TABLE B.8 Square footages of Powhatan Class tugboats.

POWHATAN CLASS T-ATF	BLAST	POWER TOOL	TOTAL SQUARE FEET
MAIN DECK	3,560	630	4,190
AFT OF PILOT HOUSE	358	358	716
UPPER FOCSLE		420	420
01 LEVEL AFT OF HOUSE	665	285	950

TABLE B.9 Square footages of Harpers Ferry Class dock landing ships (LSD).

HARPERS FERRY CLASS (LSD'S)	BLAST	POWER TOOL	TOTAL SQUARE FEET
FLIGHT DECK 725 T/D'S	16,371	2,166	18,537
FANTAIL PORT AND STARBOARD (P/S)		940	940
FOCSLE	1,907	1,271	3,178
04 LEVEL FRAME 12-21 GUN DECK		739	739
02 LEVEL FRAME 59-67		145	145
02 LADDER PLATFORM		54	54
03 LADDER PLATFORM		88	88
04 LADDER PLATFORM		130	130
(03-63-2-W) TOP OF ELEVATOR		216	216
05 LEVEL	680	519	1,199
BOAT DECK 156 T/D'S	4,689	2,010	6,699
WEATHER DECK (03-67-2-L)		234	234
P/S MAIN DECK	768	768	1,536
P/S MAIN DECK	783	783	1,566
MAIN MAST PLATFORMS		1,498	1,498
04 FORWARD RAM	742	319	1,061
05 AFT RAM	905	388	1,293
06 LEVEL WEATHER DECK	5,891	2,524	8,415
07 LEVEL	1,695	1,130	2,825

AIR MISSILE DECK		352	352
VEHICLE STORAGE TURN AREA	1,684	908	2,592
VEHICLE STORAGE FRAME 35-88 530 T/D'S	5,650	3,768	9,418
RAMP FLIGHT DECK TO P/S		900	900
RAMP TO FLIGHT DECK		260	260
RAMP P/S TO WELL		663	663
RAMP TO P/S		396	396
RAMP WIND TO TURN AREA		420	420
WIND TUNNEL 83 T/D'S	1,164	284	1,448
P/S REFUEL AT SEA STATION	265	113	378
P/S REFUEL AT STATION	245	105	350
PASSAGEWAY (01-58-2-L)		168	168
PASSAGEWAY (01-58-1-L)		96	96
LIGHT TRAP (02-18-2-L)		32	32
MOORING STATION (2-53-2-Q)		132	132
PASSAGEWAY (1-52-3-L)		112	112
STEERING GEAR ROOM		1,000	1,000
CLOSE-IN WEAPON SYSTEM (CIWS) #21 FRAME 25		389	389
CIWS #24 FRAME 55		284	284
CIWS MAGAZINE 06-57-0-M		126	126
MAST 113		232	232
MAST 122		60	60
MAST 131		504	504
MAST 140		378	378
08 LEVEL FORWARD		480	480
09 LEVEL MAST		96	96
010 LEVEL MAST		700	700
011 LEVEL MAST		671	671
08 LEVEL SIGNAL SHACK		64	64
LIFERAFT ASSM 02-59-5-W		90	90
PORT HIGH FREQUENCY TILTING WHIP ANTENNA 01-133-2-WEA		72	72
STARBOARD HIGH FREQUENCY TILTING WHIP ANTENNA 01-133-1-WEA		72	72
QUARTERDECK AREA 1-81-1 TO 1-121-1		495	495
LCAC STORAGE AREA 01-59-2-A		414	414
P/S FANTAIL FRAME 134 TO STERN EACH SIDE		552	552
ELEVATOR MACHINE ROOM		204	204
PLATFORM ABOVE REFUEL AT STATION 03-61-6-WEA		111	111
02-03-04 LEVEL BASES OF LADDERS		323	323
SPS-49 PLATFORM		976	976
SPS 67 WITH PORT YARDARMS		2,177	2,177

TABLE B.10 Square footages of USS *Denver* landing platform dock (LPD) Class ships.

USS DENVER LPD CLASS	BLAST	POWER TOOL	TOTAL SQUARE FEET
FLIGHT DECK 385 T/D'S	16,932	1,882	18,814
FORWARD OF FLIGHT DECK AREA	1,968	1,312	3,280
HANGAR BAY		613	613
FUELING STATIONS (4)		720	720
FOCSLE DECK	1,615	1,674	3,289
01 LEVEL FRAME 36-50 P/S	1,524	654	2,178
BOAT DECK 01 LEVEL	1,912	1,912	3,824
UPPER "V" STOWAGE	4,346	4,346	8,692
02 LEVEL 02-104-0-WEA	1,505	377	1,882
02 LEVEL STARBOARD BRIDGE WING		142	142
02 LEVEL PORT BRIDGE WING		142	142
03 LEVEL STARBOARD BRIDGE WING	170	212	382
03 LEVEL PORT BRIDGE WING	170	170	340
03 LEVEL 03-96-01-Q	1,281	321	1,602
04 LEVEL WEATHER DECK	3,643	2,429	6,072
05 LEVEL WEATHER DECK.		600	600
05-60-0 MAIN MAST		398	398
PORT ACCOM SPONSON		475	475
STARBOARD ACCOM SPONSON		471	471
FUEL PIT #4 PORT SIDE		732	732
05-71-1-WEA		108	108
05-71-2-WEA		108	108

TABLE B.11 Square footages of USS *New Orleans* landing platform dock (LPD) Class ships.

USS NEW ORLEANS LPD CLASS	BLAST	POWER TOOL	TOTAL SQUARE FEET
FLIGHT DECK	19,256	2,140	21,396
03 LEVEL AFT WEA 03-96-01	4,356	1,269	5,625
MAIN VEHICLE STOW RAMP		780	780
04 LEVEL FORWARD 04-34-01 WEA	6,462	718	7,180
04 LEVEL AFT 04-115-1-WEA	1,861	466	2,327
RESCUE BOAT POCKET		986	986
BOAT VALLEY PORT	577	144	721
BOAT VALLEY STARBOARD	862	215	1,077
BOAT VALLEY CENTER	2,531	633	3,164
FOSCLE	4,720	880	5,600
1-110-9-L PASSAGEWAY		1,090	1,090
01-63-1-L PASSAGEWAY		658	658
1-80-5-L P PASSAGEWAY		1,157	1,157
HANGAR BAY	2,850	900	3,750

TABLE B.12 Square footages of U.S. Coast Guard (USCG) 715 TO 726 Class ships.

USCG 715 TO 726 CLASS	BLAST	HAND	TOTAL SQUARE FEET
HELICOPTER DECK	2,102	525	2,627
HANGAR BAY	655	164	819
FOCSLE	1,346	281	1,627
FANTAIL	1,415	395	1,810
GUN DECK	728	393	1,121
P/S 01 LEVEL	1,116	1,116	2,232
P/S MAIN DECK	1,260	800	2,060

TABLE B.13 Square footages of USNS *Henry Kaiser* Class ships.

USNS HENRY KAISER CLASS T-AO 187-204	BLAST	POWER TOOL	TOTAL SQUARE FEET
HELICOPTER DECK	7,491	507	7,998
FOCSLE	10,620	3,540	14,160
01 LEVEL	18,603	7,972	26,575
TANK DECK	15,526	10,350	25,876

TABLE B.14 Square footages of Patrol Craft boats.

PATROL CRAFT BOATS	BLAST	POWER TOOL	TOTAL SQUARE FEET
WEATHER DECKS	1,231	820	2,051
FORWARD PANEL	189	126	315
AFT PANEL	142	95	237
OPEN BRIDGE	140	94	234
TOP OF PILOT HOUSE		75	75
TOP OF MID SHIP		150	150
TOP OF AFT		160	160
MAST PLATFORMS		115	115

TABLE B.15 Square footages of USNS *Rainer* Class ships.

USNS RAINER (T-AOE) CLASS	BLAST	POWER TOOL	TOTAL SQUARE FEET
FLIGHT DECK T/D'S (166)	6,460	1,615	8,075
HANGAR BAY T/D'S (42)	1,248	832	2,080

TABLE B.16 Square footages of USNS *Flint* Class ships.

USNS FLINT (T-AE 32) CLASS	BLAST	POWER TOOL	TOTAL SQUARE FEET
MAIN DECK T/D'S (250)	9,375	3,125	12,500
01 LEVEL T/D'S (350) 6"	8,925	1,575	10,500

TABLE B.17 Square footages of Landing Craft Air Cushion (LCAC) Class ships.

LCAC	BLAST	POWER TOOL	TOTAL SQUARE FEET
MAIN DECK	1477	1483	3,000

TABLE B.18 Square footages of T-AKE Dry Cargo/Ammunition Class ships.

T-AKE'S	BLAST	POWER TOOL	TOTAL SQUARE FEET
FLIGHT DECK T/D'S (1000)	9,103	1,607	10,710
HANGAR BAY T/D'S (300)	2,037	874	2,911
OUTBOARD EDGES		1,545	1,545

TABLE B.19 Square footages of T-AKE Dry Cargo/Ammunition Barge Class (YD) boats.

YD	BLAST	POWER TOOL	TOTAL SQUARE FEET
MAIN DECK	10,700	2,676	13,376
YD-256	10,380	2,595	12,975

TABLE B.20 Square footages of Arco ARDM Class ships.

ARCO ARDM-5	BLAST	POWER TOOL	TOTAL SQUARE FEET
PORT SIDE WING WALL	3,675	2,451	6,126
STARBOARD SIDE WING WALL	3,730	2,486	6,216

TABLE B.21 Square footages of Barge Class vessels.

BARGES	BLAST	POWER TOOL	TOTAL SQUARE FEET
	APL-5		
MAIN DECK	1,252	836	2,088
01 LEVEL	699	466	1,165
02 LEVEL	7,364	1,841	9,205
	YRBM-39		
MAIN DECK	1,619	405	2,024
01 LEVEL FORWARD AND AFT VESTIBULES		238	238
02 LEVEL FORWARD AND AFT VESTIBULES		238	238
03 LEVEL ROOF DECK	5,132	1,284	6,416
	YRBM-51		
MAIN DECK	926	397	1,323
01 LEVEL		360	360
02 LEVEL		638	638
03 LEVEL	1,974	494	2,468
	YRBM		
MAIN DECK	1,179	995	2,174
01 LEVEL FORWARD AND AFT		210	210
02 LEVEL FORWARD AND AFT		210	210
03 LEVEL ROOF	3,616	904	4,520
03 VESTIBULE FORWARD AND AFT		262	262
03 LEVEL SHACK ROOFS FORWARD AND AFT		868	868

TABLE B.22 Square footages of nuclear powered aircraft carrier (CVN) class ships.

CVN	BLAST	POWER TOOL	TOTAL SQUARE FEET
LANDING AREA REAGAN	47,535	2,502	50,037
ABRASIVE AREA FORWARD AND AFT OF LANDING AREA REAGAN	44,145	2,324	46,469
FAN TAIL	2,092	897	2,989
1-230-3-D		229	229
1-235-3-L		150	150
UPPER PORT BOAT DECK		446	446
UPPER STARBOARD BOAT DECK		1,508	1,508
LOWER PORT BOAT DECK		897	897
LOWER STARBOARD BOAT DECK		630	630
1-250-2-A		260	260
1-250-1-A		260	260
1-123-5-D		279	279
1-123-3-Q		72	72
3-247-1-E		416	416
08 LEVEL PORT SIDE		259	259
08-0-0-WEA		1,326	1,326
03-190-3-Q		458	458
01-250-1-WEATHER SPONSON 41 NIMITZ		504	504
02-32-4-S-RAM (ROLLING AIRFRAME MISSILE) 1 NIMITZ	805	346	1,151
03-250-5-S-RAM 2 NIMITZ	653	280	933
03-250-4-S-NSSMS (NATO SEASPARROW MISSILE SYSTEM) NIMITZ		886	886
03-34-7-S NSSMS NIMITZ		986	986
FORWARD RAM 02-32-00 PORT REAGAN	710	315	1,025
FORWARD NSSMS REAGAN	1,232	308	1,540
FRAME 66 STARBOARD WIP ANTENNA		66	66
FRAME01 PORT WIP ANTENNA		54	54
FRAME160 PT WIP ANNTENNA		48	48
FRAME198 PT WIP ANTENNA		36	36
FLIGHT DECK FRAME BOW 94-FR (G)	44,981	2,368	47,349
FLIGHT DECK FRAME 94-FR 180 (G)	31,510	1,658	33,168
FLIGHT DECK FRAME 180-FR 250 (G)	8,022	2,007	10,029
FLIGHT DECK H TO AFT OF 101 NON ABR	38,990		38,990
FLIGHT DECK 116T TO AFT TO FRAME250 ABR		44,280	44,280
ACE #1	3,200	800	4,000
ACE #2	3,200	800	4,000
ACE #3	3,200	800	4,000
ACE #4	3,200	800	4,000
HANGAR BAY #2	20,443	2,272	22,715

HANGAR BAY #3	18,796	2,088	20,884
#3 ELEVATOR WELL	1,095	273	1,368
UNREP STATION 02-240-1		541	541
UNREP STATION 01-240-7		360	360
FOSCLE 02-K-0-Q		1,9501	1,950
#2 ELEVATOR WELL	1,000	100	1,100
#4 ELEVATOR WELL	1,095	273	1,368
REFUEL AT SEA STATION 5 1-123-5-D		304	304
PASSAGEWAY 1-123-3-L		66	66
WEAPONS UNREP STATION		216	216
ANCHOR WINDLESS 01-13-0-Q		319	319
BOAT A AIRCRAFT CRANE 02-220-7		340	340
09-158-01-WEA		1,072	1,072
010-0-Q-WEA		1,239	1,239
WSC-8 SPONSON		102	102
09-96-20-L FRAME94-106 NIMITZ		340	340
AIMD STRUCTURE SHOP NIMITZ		1,085	1,085
PORT CAT WALK FORWARD		691	691
03-151-2 LIGHT LOCKER TOP		20	20
03-16-18 LIGHT LOCKER TOP		20	20
03-169-6 LIGHT LOCKER TOP		20	20
03-78-3 LIGHT LOCKER TOP		20	20
03-88-07 LIGHT LOCKER TOP		20	20
03-85-11 LIGHT LOCKER TOP		25	25
03-128-15 LIGHT LOCKER TOP		20	20
03-169-3-1 LIGHT LOCKER TOP		20	20
03-19-1-L LIGHT LOCKER DECK TOP		50	50
03-39-1-L LIGHT LOCKER DECK TOP		50	50
03-51-3-L LIGHT LOCKER DECK TOP		50	50
03-58-1-L LIGHT LOCKER DECK TOP		50	50
03-69-15-L LIGHT LOCKER DECK TOP		45	45
03-123-3-L LIGHT LOCKER DECK TOP		45	45
03-179-9-L LIGHT LOCKER DECK TOP		50	50
03-12-2 LIGHT LOCKER TOP		70	70
03-45-4 LIGHT LOCKER TOP		40	40
03-138-14 LIGHT LOCKER TOP		40	40
AFT STARBOARD TUB		55	55
CRANE TUB BY ELEVATOR #3		55	55
STARBOARD CATWALK AFT		150	150
STARBOARD GUN TUB TOP		20	20
ARMING PLATFORM		110	110
03-184-2 LIGHT LOCKER TOP		15	15

03-195-8 LIGHT LOCKER TOP		40	40
03-228-6 LIGHT LOCKER TOP		20	20
LSO PLATFORM		120	120
2 TOPS AFT OF LSO PLATFORMS		20	20
PORT CATWALK AFT		150	150
PORT GUN TUB TOP		20	20
PORT GUN TUB		125	125
STARBOARD GUN MOUNT		480	480
010 LEVEL		2,000	2,000

TABLE B.23 Square footages of Mine Countermeasures (MCM) ships

MCM	BLAST	POWER TOOL	TOTAL SQUARE FEET
01 LEVEL INCLUDING FOCSLE		3,061	3,061
02 LEVEL		1,060	1,060
02 LEVEL PORT SIDE		530	530
02 LEVEL STARBOARD SIDE		530	530
02 ½ LEVEL STARBOARD		300	300
03 LEVEL		1,764	1,764
FANTAIL		2,048	2,048
FOSCLE		1,560	1,560
01 PORT SIDE		779	779
01 STARBOARD SIDE		779	779

TABLE B.24 Square footages for the USS *Makin Island* Class ships.

MAKIN ISLAND	BLAST	POWER TOOL	TOTAL SQUARE FEET
VEHICLE RAMP FLIGHT DECK TO HB		1,152	1,152
CIWS MOUNT 22		266	266
07 LEVEL ABOVE BOAT LAUNCH	525	225	750

TABLE B.25 Square footages of the Freedom littoral combat ship (LCS).

FREEDOM LCS	BLAST	POWER TOOL	TOTAL SQUARE FEET
FLIGHT DECK 286 6"	4,968	552	5,520
FOSCLE	4,617	513	5,130
BOAT DECK PORT		434	434
HANGAR BAY 154 6"	3,696	411	4,107

Index

Page numbers followed by *f* indicate figures, those followed by *t* indicate tables.

www.ingramcontent.com/pod-product-compliance
Lightning Source LLC
LaVergne TN
LVHW060041210926
848481LV00013B/3922
* 9 7 8 1 5 7 5 9 0 4 9 3 1 *